E FOR ADDITIVES

An indispensable reference book which cracks the 'E'
Numbers code, to help you make better-informed decisions
about the foods you buy and eat.

'Leg of spider, ear of toad, wing of bat, permitted
flavouring and artificial colouring!'

E

FOR ADDITIVES

The Complete 'E' Number Guide

Compiled and Written by

Maurice Hanssen
with
Jill Marsden

THORSONS PUBLISHERS LIMITED
Wellingborough, Northamptonshire

First published October 1984
Second Impression November 1984
Third Impression December 1984
Fourth Impression January 1985
Fifth Impression February 1985
Sixth Impression February 1985
Seventh Impression March 1985
Eighth Impression March 1985
Ninth Impression July 1985
Tenth Impression August 1985

British Library Cataloguing in Publication Data

Hanssen, Maurice
    E for additives.
    1. Food additives — Tables
    I. Title
    664'.06'0212   TX553.A3

    ISBN 0-7225-1150-7

Printed and Bound in Great Britain

Contents

Acknowledgements

This book is dedicated to those who helped make it possible:

To my Research Assistant, Jill Marsden B.Sc., whose strong will and tenacity were needed over many arduous months in our attempt to produce daylight where there was darkness.

To Elizabeth Brown and Maureen Sharp whose typing and re-typing of difficult material made my publisher's lot a happy one.

To Leslie Kenton who wrote such a thoughtful Foreword. But those who enjoy her articles in *Harpers and Queen*, and her fine books, would expect that!

Foreword

This comprehensive book is one which in a sense I wish need never have been written. I would prefer to live in a world where we harvested our foods fresh from the earth, ate them immediately and never had to give a thought to food preservatives, artificial emulsifiers and stabilizers, anti-oxidants and permitted colours. Alas, we do not live in such a world. High technology food production and elaborate chains of food distribution have created a situation in which food additives are necessary. Yet for the protection of oneself and one's family it is also necessary to be well informed about these hundreds of additives in quite specific terms and highly aware of the possible implications of their inclusion in our daily diet.

I therefore welcome Maurice Hanssen's *E For Additives*. Mr Hanssen has produced a simple-to-follow yet remarkably ambitious guide which can help people make informed decisions about the foods on their supermarket shelves even before they buy them. He carefully explains both the pros and cons of food additives, clarifies the meaning of such commonly used but little understood words as 'stabilizers' and 'tenderizers', and offers a quick-to-use guide to each specific additive, its name, where it comes from, the possible adverse effects of using it and a list of typical products in which it is used. This book is a useful tool for anyone concerned about the health of himself and his family. I for one would not want to be without it.

LESLIE KENTON

Introduction

What are 'E' Numbers?
Although it is not very obvious when you go through Customs, a stated objective of the European Economic Community or Common Market is to harmonize laws.

So that foods can be moved from country to country within the Common Market, a list of additives that are generally recognized as safe has been introduced. These are the 'E' numbers. Additions are made to the list from time to time as new substances are found to be suitable for use in foods. Substances are removed from the list if they are found to present problems or if they can be replaced by a more effective substance with greater safety.

Are Additives Necessary?
Sugar and salt are perhaps the most common food additives and are very important in the preservation of foods. But excessive sugar and salt is not healthy. There is a likely relationship between too much salt and high blood pressure and between excessive sugar and dental decay as well as overweight. So their use is valuable but their consumption needs to be moderated. The same is true of most other additives — they are tested for safety — but if you read your labels carefully you can control your intake.

Quite a few of the 'E' numbers are of substances that occur in nature such as vitamin C and lecithin.

This book is to help you to decide whether the additives in your food are the sort that are essential to keep it in good condition, such as when the preservation of meat products is far more important than the risks of possibly fatal food poisoning from eating tainted meat, whether it is just to assist the manufacturing processes which produce the food, or whether it is to colour, to enhance the flavour or even to allow the manufacturer to add extra water or to hide excess fat without the awareness of the consumer.

The introduction of 'E' numbers has given us, once the code can be cracked, more detail than we have ever before been given over exactly what it is we are eating.

We have for the first time enough information appearing on labels to enable us to choose whether the additives are the sort that we are happy to use or would prefer to do without. Remember too, that although perhaps one in ten will find the yellow colour tartrazine does not suit them, nine out of ten will find that it does! Question whether the additives are just for the convenience of the manufacturer, enabling him to use lower quality raw materials or processing methods without the knowledge of the consumer, or whether they enable the food to be brought to you better tasting and in better condition than would be possible without the use of additives. The E.E.C. has been issuing Directives on additives since 1962, so learn to read the label wisely.

All foods made after 1 January 1986 will (except for flavourings) have to have the 'E' number or the actual name in the list of ingredients. Until then is a transitional period to enable manufacturers to change their packaging. The old sort of uninformative labelling with words such as 'permitted colouring' will then cease. The very word 'permitted' whilst giving a feeling of security, means very little for if the manufacturer used a colour that was *not* permitted then he would be breaking the law and, incredibly, there are no regulations covering the composition of flavours except the general requirement that a food should be safe. So there is no such thing as a permitted flavour — a serious gap in the protection of the consumer.

There would not be enough room on many products to spell out all the additives used in full without making some packs look like a food chemist's dictionary so the 'E' number idea can be very useful to the consumer especially if he or she has a sensitivity to certain additives or a desire to choose what goes into the food.

Also included in this book are numbers without the prefix 'E'. These are, at the time of writing, proposals and need not necessarily be adopted. Also some 'E' numbers are not at present permitted in Britain but are also included for the sake of completeness.

How to Use This Book

Some additives have several different uses but in general 'E' numbers follow a numerical system according to their main function. The first part of this book is laid out numerically, according to the 'E' number, so that you can quickly look up the information on the label.

Each entry lists the number, the name of the substance, its uses, any reported adverse effects, whether or not it occurs naturally and, from time to time, notes for the consumer. At the end is an alphabetical list of 'E' numbers so that any ingredient given by name instead of number can be checked out.

The labelling regulations state that ingredients, including water, must be given in descending order by weight so it is a useful indicator of the amount of the additive present to see, for example, whether it comes above or below salt.

The Ministry of Agriculture, Fisheries and Food produces a little guide to 'E' numbers which does not give any indication of what the substances are and what they do. In this book an attempt has been made to gather as much information as possible and there will inevitably be some sins of omission and indeed the E.E.C. will certainly make continual changes in their proposals. Any changes which are brought to our attention will be rectified, where appropriate, in subsequent editions. We would also be glad to know of products using additives for which we found no typical examples.

Information for the Consumer

Apart from the list of ingredients in descending order by weight and with the warning to look out for that magic ingredient water especially in products such as ham where you would not normally wish to see it, there are many safeguards for the consumer who is aware of what are the rules of the game.

Date marking is now required on most pre-packed foods (with a few exceptions, such as frozen foods, wine and vinegar) unless they have a shelf-life of at least 18 months. Even products with a very long shelf-life may be marked, but this is not mandatory. This shall be expressed as *either:*

- A best before date (day, month, year) plus storage conditions (if necessary).

Or:

- If the food has a life of between 3 months and 18 months, a best before end date (month, year).
- If the food has a 'life' of between 6 weeks and 3 months, a best before date (day, month) plus storage conditions (if necessary).
- If the food is perishable and is intended for consumption within 6 weeks of being packed, a sell by date (day, month) plus storage conditions and a storage period after purchase.

There is no reason why you should not buy overdue products, especially if they are reduced in price, because the onus is on the shopkeeper to provide goods which live up to the quality of their description, in other words they must not be bad or 'off'. With the longer time datings you are safe in buying goods that are near the end of their expiry date if the shop is clean and well maintained. However if such a product has deteriorated, even if bought at a special price, your legal rights are not affected and you should complain first of all to the shop manager then, if no satisfaction is obtained, to your local Trading Standards Officer who you can locate through the Town Hall. It is often preferable, though, to write a nice letter, fully documented, with a sample, to the Managing Director of the company concerned who will often, for the sake of goodwill (and most of the food companies are very jealous of their good reputation), refund your cost and may even give you something extra besides. However, if you are on the make, beware, because most manufacturers keep very accurate records of complainants and get wise to the person who frequently finds a dead mouse in the meat pie.

Foods for special nutritional purposes are subject to the provisions of an EEC Directive which strictly controls all claims and declarations in respect of infant, diabetic, slimming and other foods which make them appear that they are for a group of people with special nu... tional needs. There is a problem in that some excellent foods which have a nutritional purpose may not, in the future, be able to declare it without a Medicines Licence! For example, a bran based breakfast cereal may not be able to say that it helps prevent constipation even though the Roman Army almost 2,000 years ago had to eat wholemeal bread to keep them free of constipation and no serious authority has doubted the benefits of bran ever since. Too often we are seeing legislation which is designed for consumer protection which effectively shields the consumer from the information needed to make an informed decision. It should surely be sufficient that advertising is decent, honest and truthful.

Hyperactivity in Children

Hyperactive children bring much strain and exhaustion to parents who have to manage offspring that sleep only a few hours out of twenty-four, may suffer from eczema and asthma and can not be calmed down.

As they grow older they become even more active and can easily become hurt. Difficulties are experienced with speech, balance and learning even if the IQ is high. They suffer from excessive thirst and often are prone to respiratory difficulties.

It was to help such parents and children that the Hyperactive Children's Support Group was formed in 1977. It is now a registered charity. The Secretary is Mrs Sally Bunday, 59 Meadowside, Angmering, West Sussex, BN16 4BW. (Please enclose an S.A.E.)

The Group recommends that parents try a diet based on the work of the American doctor, Ben Feingold. First, this means cutting out all food and drink containing synthetic colours or flavours, avoiding glutamates, nitrites, nitrates, BHA, BHT and benzoic acid. Second, for the first four to six weeks, foods containing natural salicylates (like aspirin chemically) should be avoided and then re-introduced *one at a time* to see if they cause problems. Such foods include almonds, apples, apricots, peaches, plums, prunes, oranges, tomatoes, tangerines, cucumbers, most soft fruits, cherries, grapes and raisins.

The 'E' numbers of additives that the HACSG recommend should be avoided are:

E102   Tartrazine   E110   Sunset Yellow FCF

E104	Quinoline Yellow	E120	Cochineal
107	Yellow 2G	E122	Carmoisine
E123	Amaranth	155	Brown HT
E124	Ponceau 4R	E210	Benzoic acid
E127	Erythrosine	E211	Sodium benzoate
128	Red 2G	E220	Sulphur dioxide
E132	Indigo Carmine	E250	Sodium nitrite
E133	Brilliant blue FCF	E251	Sodium nitrate
E150	Caramel	E320	Butylated hydroxyanisole
E151	Black PN	E321	Butylated hydroxytoluene
154	Brown FK		

Additives which are either dangerous to asthmatics or aspirin-sensitive people, and could reasonably be added to the HACSG listing or not allowed in food intended for babies or young children are:

E212	Potassium benzoate
E213	Calcium benzoate
E214	Ethyl 4-hydroxybenzoate
E215	Ethyl 4-hydroxybenzoate, sodium salt
E216	Propyl 4-hydroxybenzoate
E217	Propyl 4-hydroxybenzoate, sodium salt
E218	Methyl 4-hydroxybenzoate
E219	Methyl 4-hydroxybenzoate, sodium salt
E310	Propyl gallate
E311	Octyl gallate
E312	Dodecyl gallate
621	Sodium hydrogen L-glutamate
622	Potassium hydrogen L-glutamate
623	Calcium dihydrogen di-L-glutamate
627	Guanosine 5'-(disodium phosphate)
631	Inosine 5'-(disodium phosphate)
635	Sodium 5'-ribonucleotide

In *The Lancet* (Editorial, 20 March 1982, p. 662) the view was taken that most clinicians will say that the evidence 'does not yet justify prescribing the treatment, but does justify supporting families who have decided to try it themselves.'

The 'E' Number Categories

Permitted Colours (E100-E180)
Most colours are used for cosmetic reasons — to make the product look attractive. Quite a lot are natural in origin. For example, chlorophyll, the green colour of plants is the only permitted colour for sage cheese.

Some additional colours are being considered by the E.E.C. for an 'E' prefix, but the serial numbers which they have been allocated do not appear on food labels at the moment. These are 101a, 107, 128, 133, 154, 155.

The number of permitted colours is likely to be reduced as there is much dispute over the safety of quite a few — especially certain black, brown, yellow, green and red colours.

Some additional natural substances or extracts which also have a use as a colour are permitted but do not have E numbers. These include saffron, turmeric root, sandalwood and paprika. Citrus fruits can be marked with methyl violet.

Apart from marking, fresh meats, poultry, fish, fruits and vegetables may not be coloured. Neither can dried or condensed milk, tea and coffee.

Preservatives (E200-E290)
Preservatives are additives which check or prevent the growth of micro-organisms. Therefore from the point of view of providing safe food to the consumer at a point distant from the place of manufacture they constitute a group of substances which, on the whole, are far safer than mouldy food.

The ancient Greeks burnt sulphur over the wine in the cask before sealing, thus producing the preservative sulphur dioxide which is used in wines and many other foods to this day.

In the ideal world we would probably be better off with fresh, natural produce, but as things stand the moderate use of preservatives allows a variety in our diets which has benefits which may outweigh the risks. Such methods of preservation as alcohol, salting, adding sugar, smoking and adding herbs or spices are not included in this category.

Avoid preservatives only when you are sure that the food is fresh.

A number of the substances in the group have other functions, e.g., vinegars (acetic acids) are used for flavouring and carbon dioxide as a propellent in aerosols. Some of the additives are also classed as miscellaneous because they have additional functions. These include E260, E261, 262, E262 and E290.

Permitted Anti-oxidants (E300-E321)
The oxygen in the air causes many food products to undergo changes which make them unfit to eat causing, for example, rancidity in oils and fats.

The anti-oxidants vary from natural or nature-identical substances such as vitamin C (ascorbic acid) and vitamin E (the tocopherols) to others, the safety record of which is at least debatable such as E320, BHA and E321 BHT.

The amounts that may be used are controlled at various levels in different foodstuffs depending upon the amount needed to be effective.

Without anti-oxidants it would be impossible to provide many of the ready-packed foods on the supermarket shelves.

An additional anti-oxidant, ethoxyquin which does not have an 'E' number, is permitted for preserving apples and pears by being used as a surface spray to prevent 'scald'.

Emulsifiers and Stabilizers: Some numbers between E322 and 494.
When an egg is used to make a mayonnaise or to bind a sauce the lecithin in the egg works as an emulsifier. Lecithin is E322 which nowadays, for reasons of economy, is usually produced from soya beans.

Many of the emulsifiers and stabilizers are safe and natural. The polyphosphates, E450, which are used to 'tenderize' many cured meats including ham are in reality a way of adding that low cost substance water. Frozen poultry often also has additions of polyphosphates enabling substantial increases in weight to be made.

In July 1979 the magazine of the Consumer Association, *Which?*, did a survey on cooked ham and found water added in percentages ranging from nil right up to 42 per cent in certain canned products. They found no relation between price and water content.

Two additives are permitted which do not have proposed or actual 'E' numbers and these are dioctyl sodium sulphosuccinate and the extract of Quillaia which is allowed in soft drinks.

Sweeteners (E420-421)

Only two of the permitted sweeteners have 'E' numbers, sorbitol and mannitol.

The regulations in the U.K. concerning permitted sweeteners were revised in 1983 when four exhaustively tested new sweeteners were added which are acesulfame potassium, aspartame, thaumatin and xylitol. Hydrogenated glucose syrup, isomalt and certain forms of saccharin are also listed as permitted sweeteners.

Other sweet substances are classed as foods rather than permitted sweeteners and these include sugar (sucrose), glucose (dextrose), fruit sugar (fructose) and milk sugar (lactose).

Solvents (E422)

Certain liquids which are not natural food substances are used to extract or dissolve foods so that they can be incorporated into products. A commonly used substance is alcohol in the form of ethyl alcohol or ethanol with which many flavouring and colouring extracts are made.

Other permitted solvents include ethyl acetate, ether, various glycerols (glycerol is the only solvent with an 'E' number). The other two which may be used are iso-propyl alcohol and propylene glycol.

Mineral Hydrocarbons (905-907)

Mineral hydrocarbons are used to prevent drying out of certain foods and as a polish to make foods shiny.

They are permitted in dried fruits, sugar confectionery, chewing sweets, cheese rind and eggs which must be marked 'sealed'.

Modified Starches (E1400-E1442)

These are to be found on certain products (we found E1422 on several dairy products such as milkshakes in the Summer of 1984) but it must be *emphasized emphatically* that these E numbers were proposals that were never accepted and are definitely *not* permissible to be used on packs in the U.K., even though there is no suggestion that they are not safe.

So that you know what you are getting if one of these numbers crops up, the full listing is as follows:

E1400	White or yellow dextrins, roasted starch.
E1401	Acid treated starches.
E1402	Alkaline treated starches.
E1403	Bleached starches.
E1404	Oxidised starches.
E1410	Mono starch phosphate.
E1411	Di starch phosphate. } produced in different ways.
E1412	Di starch phosphate.
E1413	Phosphated di starch phosphate.
E1414	Acetylated di starch phosphate.
E1420	Starch acetate. } produced in different ways.
E1421	Starch acetate.
E1422	Acetylated di starch adipate.
E1423	Acetylated di starch glycerol.
E1430	Di starch glycerol.
E1440	Hydroxypropyl starch.
E1441	Hydroxypropyl di starch glycerol.
E1442	Hydroxypropyl di starch phosphate.

Miscellaneous Additives (E170-927)

This group is not a numerical run between these two numbers. So many different substances with various purposes not covered by other regulations are incorporated in the Miscellaneous section that no useful guidance can be given that is not better found under the individual listing.

In this group is found substances like monosodium glutamate which is a flavour modifier or enhancer, propellant gases for food aerosols, bulking agents, anti-foaming substances, acids and glazing and releasing agents.

620, 622 and 623 are allowed in dietetic foods; 903 in chocolate and sugar confectionery; 353 in wine, 385 is allowable in canned fish and shellfish and brandy.

Certain additives (see text) are not permitted in baby and infant food and neither is 2-aminoethanol which may be used in peeled fruit and peeled vegetables so particular care needs to be taken when buying these to be sure that the substance is not included if they are to be eaten by little children.

Meat Products

Regulations governing meat products and spreadable fish products were laid before Parliament in October 1984. Like the E numbering provisions, these will come into full operation by July 1986.

Polyphosphates (E450) allow the manufacturer to add water to meat products without it becoming obvious to the consumer. If the meat is cooked or raw and contains added water, then the producer will have to declare: 'with not more than x per cent added water.' X is the maximum added water content of the food. On the other hand, if the meat is uncooked and cured, such as bacon, of which more than 10 per cent is added water, then the declaration has to say 'with not more than y per cent added water; but that does *not* mean that this figure represents the amount of added water — y represents a multiple of 5 by which the percentage of water in the product exceeds 10 per cent! Finally, to make matters clear to our (presumably computer owning!) consumer — if it is cooked pure meat then the declaration has to say 'with not more than z per cent added water', z being an indication in multiples of 5 of the percentage of water added.

There is a list of parts of the carcass which may *not* be used in *uncooked* meat products — and may therefore be used in cooked meat products. You will be glad to know what comprehensive use manufacturers of cooked meat products can make of the slaughtered animal because they can use the brains, foot, large intestine, small intestine, lungs, oesophagus, rectum, spinal cord, spleen, stomach, testicles and udder. There has to be a argument for manufacturers to tell us just what parts of the animal are used and how much, not just the blanket description 'offal'.

A meat pie weighing between 100g and 200g must have a meat content of not less than 21 per cent of the total. If the pie weighs less than 100g the meat content can shrink to 19 per cent of the food, otherwise the meat content can soar to the dizzy heights of 25 per cent as a minimum *but*, of these percentages, the lean meat content need only be half so, at the worst, a quarter of a pound pork pie may contain just over a third of an ounce of lean meat — and it may include unexpected parts of the beast.

The true nature of the contents are then disguised in taste and appearance by the use of flavour enhancers, such as monosodium glutamate (number 621). It can then be coloured, flavoured and, after the addition of the appropriate amount of water, you can have, at the worst, a very fatty pie but one which looks and tastes good. Though of course there are many pie manufacturers who certainly do use the finest ingredients, it would be worth their while making clear claims. The fat content of burgers and sausages is also controlled, in general so that the fat content of the meat part of the product does not exceed 35 per cent.

Many German meat products are labelled with their fat content. In order to be able to eat sensibly we should demand that such information be available throughout the E.E.C.

Permitted Colours

E100   **Curcumin (C.I. 75300)**

Origin   Extract of turmeric root.

Function   Orange-yellow colour for whole food, or surface only.

Adverse   None known.
Effects

Typical   Savoury rice
Products   Curry powders
            Margarine
            Processed cheese

E101   **Riboflavin (Lactoflavin; Vitamin B2)**

Origin   Produced from yeast, or more usually synthetically. Occurs naturally in liver, kidneys, green vegetables, eggs and milk and a small amount is synthesized by bacteria in the large intestine.

Function   Yellow or orange-yellow colour; vitamin B_2.

Adverse   Little riboflavin is stored in the body; excess of
Effects   requirements is excreted in the urine. No toxic problems.

Typical   Processed cheeses
Products

101(a)   **Riboflavin – 5'-phosphate (Riboflavin – 5'-[Sodium phosphate])**

Origin   Prepared by chemical action on riboflavine.

Function   Yellow colour; vitamin B_2.

Adverse   None known.
Effects

Typical   Various sugar products
Products   Jams

                  Under consideration by
                     the E.E.C. for an 'E'
                     prefix.

E102   **Tartrazine (C.I. 19140)**

Origin   Synthetic, an azo dye (see glossary).

Function   Yellow colour.

Adverse   Susceptible people, especially those sensitive to
Effects   aspirin, and asthmatics, are sensitive to tartrazine.
Reactions include urticaria (skin rashes), rhinitis
(hayfever), bronchospasm (breathing problems),
blurred vision and purple patches on the skin.
It has recently been suggested that tartrazine in
fruit cordials may be responsible for wakefulness
in small children at night. See also pages 12-14.

Typical   Packet convenience
Products   foods
Rind of cheese
Smoked cod and
    haddock
Chewing gum
Sweets
Lime and lemon squash
Seafood dressing
Mint sauce and jelly
Packet dessert topping
Tinned fruit pie filling
Tinned processed peas
Salad cream
Prepacked cakes
Marzipan
Piccalilli
Brown sauce
Maple flavour syrup
Fizzy drinks
Shells of capsules

A very commonly used colour.

E104   **Quinoline Yellow (C.I. 47005)**

Origin   Synthetic 'coal tar' dye (see glossary).

Function   Dull yellow to greenish yellow colour.

Adverse   See pages 12-14.
Effects

Typical   Scotch eggs
Products   Smoked haddock

---

107   **Yellow 2G**

Origin   Synthetic 'coal tar' dye, and azo dye (see glossary).

Function   Food colour.

Adverse   Yellow 2G belongs to a group of chemical dyes
Effects   known as azo dyes. People who suffer from
                asthma and those sensitive to aspirin may also
                show an allergic reaction to this colour. See also
                pages 12-14.

Typical   —
Products   Under consideration by
                  the E.E.C. for an 'E'
                  prefix.

E110   **Sunset Yellow FCF (C.I. 15985)**

Origin   Synthetic 'coal tar' dye, and azo dye (see glossary).

Function   Yellow colour.

Adverse   An azo dye to which some people have an allergic
Effects   reaction. Important risk of allergy especially in
                  people showing aspirin sensitivity, producing
                  urticaria (skin rash), angioedema (swelling of the
                  blood vessels), gastric upset and vomiting. See
                  also pages 12-14.

Typical   Hot chocolate mix
Products   Packet soup
                  Sweets
                  Packet trifle mix
                  Yogurt whip
                  Packet sorbet mix
                  Orange jelly biscuits
                  Packet breadcrumbs
                  Packet cheese sauce
                      mix
                  Orange squash
                  Marzipan
                  Swiss roll
                  Apricot jam
                  Lemon curd

E120 **Cochineal (Carmine of Cochineal; Carminic acid; C.I. 75470)**

Origin Cochineal is the natural red colour from the egg yolks and the fatty parts of the dried female insect *Dactilopius coccus* (*Dactilopiidae*) a native of central America, and the Canary Islands. It contains about 10 per cent of carminic acid. Carmine is produced from cochineal, as its aluminium lake.

Function Red colour.

Adverse Effects See pages 12-14.

Typical Products Now used fairly rarely because of high cost, but available as cochineal food colour for home cooking. Largely replaced by E124 in manufacturing.

E122   **Carmoisine (Azorubine; C.I. 14720)**

Origin   Synthetic azo dye (see glossary).

Function   Red colour.

Adverse Effects   An azo dye, therefore producing adverse reactions in sensitive people, or people with aspirin allergy, or asthmatics. These reactions may include urticaria (skin rashes) or oedema (swelling). See also pages 12-14.

Typical Products

Packet soup mix
Blancmange
Packet breadcrumbs
Packet jellies
Sweets
Packet cheesecake mix
Brown sauce
Savoury convenience
    food mix
Prepacked Swiss roll
Prepacked sponge
    pudding
Marzipan

E123   **Amaranth (C.I. 16185)**

Origin   Synthetic 'coal tar' dye and azo dye (see glossary).

Function   Red colour.

Adverse Effects   An azo dye, therefore to be avoided by people with aspirin sensitivity as it may cause urticaria (skin rash). See pages 12-14.

Typical *Products*	Packet soup Packet cake mix Packet trifle mix Liquid vitamin C preparations Gravy granules Tinned fruit pie fillings Quick setting jelly mix

---

E124   **Ponceau 4R (C.I. 16255)**

Origin   Synthetic 'coal tar' dye, and azo dye (see glossary).

Function   Red colour.

Adverse   An azo dye, so should be avoided by people with
Effects   aspirin sensitivity, and asthmatics. See pages 12-14.

Typical
Products
	Packet trifle mix
	Packet cheesecake mix
	Packet cake mix
	Packet soup
	Seafood dressing
	Dessert topping
	Tinned strawberries
	Tinned cherry,
	   redcurrant and
	   raspberry pie fillings
	Quick setting jelly mix

E127   **Erythrosine (C.I. 45430)**

Origin   Synthetic 'coal tar' dye (see glossary).

Function   Red colour. Also used in disclosing tablets for revealing plaque on teeth.

Adverse Effects   Can cause photoxicity (sensitivity to light). See also pages 12-14.

Erythrosine contains 577mg of iodine per gram and consumption of considerable amounts of foods containing erythrosine could increase the circulating thyroid-hormone to levels high enough to cause hyperthyroidism (overactive thryoid).

Typical Products
Glacé cherries
Scotch eggs
Tinned red cherries,
    strawberries and
    rhubarb
Packet trifle mix
Quick custard mix
Biscuits
Prepacked Swiss roll

---

128   **Red 2G (C.I. 18050)**

Origin   Synthetic 'coal tar' dye, and azo dye (see glossary).

Function   Red colour.

Adverse   Further toxicological studies required. See pages
Effects   12-14.

Typical   Sausages
Products   Cooked meat products

                  Under consideration by
                      the E.E.C. for an 'E'
                      prefix.

E131   **Patent Blue V (C.I. 42051)**

Origin   Synthetic 'coal tar' dye (see glossary).

Function   Dark bluish-violet colour and diagnostic agent,
                  used to colour the lymph vessels.

Adverse   To be avoided by patients with a history of allergy.
Effects   Allergic reactions may occur immediately or after
                  a few minutes. They consist of skin sensitivity,
                  itching and urticaria (nettle rash). More severe
                  reactions, including shock and breathing
                  problems, occur rarely. Nausea, low blood-
                  pressure and tremor have been reported. See also
                  pages 12-14.

Typical   Scotch eggs
Products

E132   **Indigo Carmine (Indigotine; C.I. 73015)**

Origin   Synthetic 'coal tar' dye (see glossary).

Function   Blue colour and diagnostic agent (used to test whether the kidneys are functioning normally by producing blue urine after Indigo Carmine is injected into veins or muscles).

Adverse Effects   People with a history of allergy should avoid this colour. May cause nausea, vomiting, high blood-pressure, hypertension and occasionally allergic reactions such as skin rash, pruritus (itching) and breathing problems. See also pages 12-14.

Typical Products   Blancmange
Biscuits
Sweets
Savoury convenience
    food mix

133   **Brilliant Blue FCF (C.I. 42090)**

Origin   Synthetic 'coal tar' dye (see glossary).

Function   Blue colour which can produce green hues in combination with tartrazine.

Adverse Effects   See pages 12-14.

Typical Products   Tinned processed peas

Under consideration by
    the E.E.C. for an 'E'
    prefix.

E140   Chlorophyll (C.I. 75810)

Origin   Pure chlorophyll is not easy to isolate and the chlorophyll which is commercially available contains other plant pigments, fatty acids and phosphatides, and is known as 'technical chlorophyll'. The usual sources are nettles, grass and lucerne.

Function   Food colour (green) and for medicines.

Adverse Effects   None known.

Typical Products   Fats
Oils
Soaps
Naturally green
    vegetables and fruits
    preserved in a liquid

E141   Copper complexes of Chlorophyll and Chlorophyllins (C.I. 75810; Copper phaeophytins)

Origin   Derived from chlorophyll by substitution of groups of atoms within the chlorophyll molecule.

Function   The copper complexes are olive-green oil-soluble colours; the chlorophyllins are green water-soluble colours.

Adverse Effects   None known.

Typical Products   Green vegetables and
    fruits preserved in a
    liquid

E142   **Green S (Acid Brilliant Green BS; Lissamine Green; C.I. 44090)**

Origin   Synthetic 'coal tar' dye (see glossary).

Function   Green colour.

Adverse Effects   None known.

Typical Products   Packet cheesecake mix
Tinned peas
Packet breadcrumbs
Gravy granules
Mint jelly and sauce

E150   **Caramel**

Origin   By the action of heat or chemicals on carbohydrates.

Function   Brown colour; flavouring agent.

Adverse   The safety of caramel has long been questioned.
Effects   The number of types now available has been reduced to six to meet all the needs of the food industry and work is being carried out to find the safest form. Caramel produced with ammonia has been shown to cause vitamin B_6 deficiency in rats. (Spector and Huntoon, *Toxicology and Appl. Pharm.*, 62, 172-178, 1982. *Nutrition Bulletin* 1982.)

Typical   Chocolate dessert whip
Products   Oyster sauce
Biscuits
Packet soup
Sachet marinade
Tinned sauce
Scotch eggs
Packet cake mix
Savoury convenience
    food mix
Sweet pickle
Pickled onions
Gravy granules
Mint jelly
Prepacked cakes
Soya sauce
Fruit sauce

E151   **Black PN (Brilliant Black BN; C.I. 28440)**

Origin   Synthetic 'coal tar' dye, and azo dye (see glossary).

Function   Black colour.

Adverse Effects   Intestinal cysts were found in pigs given black PN in a 90-day feeding study. Further studies required. See also pages 12-14.

Typical Products   Blackcurrant
     cheesecake mix
Brown sauce

E153   **Carbon Black (Vegetable Carbon)**

Origin   Natural — from burning of plant material.

Function   Black colour.

Adverse Effects   Probably few risks when the method of manufacture is perfected and a plentiful supply of oxygen is available for the combustion, but in the United States it is banned in the belief that it may be implicated in causing cancer.

Typical Products   Concentrated fruit
     juices
Jams
Jellies

154   **Brown FK (Kipper Brown; Food Brown)**

Origin   Synthetic mixture of azo dyes (see glossary).

Function   Brown colour, especially for kippers.

Adverse Effects   Experiments with bacteria have shown that two of the colour's constituents cause genetic mutation. See also pages 12-14.

Typical Products   Kippers
Smoked mackerel

Under consideration by
   the E.E.C. for an 'E'
   prefix

155   **Brown HT (C.I. 20285)**

Origin   Synthetic 'coal tar' dye and azo dye (see glossary).

Function   Brown colour.

Adverse Effects   An azo dye, therefore to be avoided by people with asthma, aspirin sensitivity and skin sensitivity. See also pages 12-14.

Typical Products   —

Under consideration by
   the EEC for an 'E'
   prefix.

E160(a) **alpha-carotene, beta-carotene, gamma-carotene (C.I. 75130)**

Origin Extracts of natural plant pigments found especially in carrots, green leafy vegetables, tomatoes, apricots, rosehips and oranges.

Function Orange-yellow colour; becomes vitamin A in the body.

Adverse Effects None known.

Typical Products
Soft margarine
Butter/margarine
Yogurt dessert whip
Prepacked coffee
   sponge cake
Sandwich cake

E160(b) **Annatto, Bixin, Norbixin (C.I. 75120)**

Origin A vegetable dye from the seed coats of the tropical Annatto tree (*Bixa orellana*).

Function Yellow to peach colour.

Adverse Effects None known.

Typical Products

Margarine
Cheshire cheese
Butter — especially to
    darken pale-coloured
    butter produced in
    the winter
Frying oil
Carton coleslaw
Prepacked sponge
    pudding

E160(c)   Capsanthin (Capsorubin)

Origin   Natural extract from paprika.

Function   Flavouring; orange colour.

Adverse Effects   None known.

Typical Products   Processed cheese slices

E160(d)   Lycopene (C.I. 75125)

Origin   Natural plant extract from tomatoes.

Function   Red colour.

Adverse Effects   None known.

Typical Products   —

E160(e)   **beta-apo-8'-carotenal (C_{30}) (beta-8'-apocarotenal)**

Origin   Natural plant derivative.

Function   Orange to yellowish red colour.

Adverse Effects   None known.

Typical Products   —

E160(f)   **Ethyl ester of beta apo-8'-carotenoic acid (C_{30})**

Origin   Natural plant derivative.

Function   Natural orange to yellow colour.

Adverse Effects   None known.

Typical Products   —

E161
E161(a)
Xanthophylls
Flavoxanthin (C.I. 75135)

Origin   A carotenoid pigment. Carotene is one of the
plant pigments of green leaves isolated from a
species of buttercup (*Ranunculus acris*).

Function   Yellow colour.

Adverse   None known.
Effects

Typical   —
Products

E161
E161(b)
Xanthophylls
Lutein (C.I. 75135)

Origin   Related to carotene, one of the plant pigments
present in abundance in green leaves. Also present
in egg yolks.

Function   Yellow to reddish colour.

Adverse   None known.
Effects

Typical   —
Products

E161	**Xanthophylls**
E161(c)	**Cryptoxanthin (C.I. 75135)**

Origin Related to carotene, part of the green plant pigment chlorophyll, especially well represented in the petals and berries of the *Physalis* (Bladder Cherry, Cape Gooseberry) genus (*Solanaceae*, the potato and tomato family) and also present in orange rind, egg yolk and butter.

Function Yellow colour.

Adverse Effects None known.

Typical Products —

E161	**Xanthophylls**
E161(d)	**Rubixanthin (C.I. 75135)**

Origin Related to carotene, especially present in rosehips.

Function Yellow colour.

Adverse Effects None known.

Typical Products —

E161   **Xanthophylls**
E161(e) **Violoxanthin (C.I. 75135)**

Origin   Natural extract from the plant pigment carotene,
            especially isolated from yellow pansies (*Viola
            tricolor*).

Function   Yellow colour.

Adverse   None known.
Effects

Typical   —
Products

E161   **Xanthophylls**
E161(f) **Rhodoxanthin (C.I. 75135)**

Origin   A naturally occurring carotenoid pigment found
            in small amounts only in, for example, the seeds
            of the Yew tree (*Taxus baccata*).

Function   Yellow colour.

Adverse   None known.
Effects

Typical   —
Products

E161	**Xanthophylls**
E161(g)	**Canthaxanthin (C.I. 75135)**

Origin   A widely-distributed carotenoid pigment which can be isolated from some mushrooms and flamingo feathers.

Function   Natural orange colour.

Adverse Effects   None known.

Typical Products
Mallow biscuits
'Sun-tan' capsules
   which when taken
   make the skin yellow
   simulating a tan but
   without sun-
   screening effect.

E162   **Beetroot Red (Betanin)**

Origin   Natural extract of beetroot.

Function   Deep purplish-red colour.

Adverse Effects   None known.

Typical Products   Oxtail soup.

E163   **Anthocyanins (Schultz 1394 & 1400)**

Origin   Natural plant pigments — E163(a) Cyanidin (red); E163(b) Delphinidin (blue); E163(c) Malvidin (purple); E163(d) Pelargonidin (red brown); E163(e) Peonidin (dark red); E163(f) Petunidin (dark red). They are all red, blue or violet pigments which are present in the cell sap of many flowers, fruits, stems and leaves.

Function   Food colouring.

Adverse   None known.
Effects

Typical   —
Products

E170   **Calcium carbonate (Chalk; C.I. 77220)**

Origin   Naturally occurring mineral.

Function   Alkali, firming agent, release agent (in vitamin tablets), calcium supplements and surface food colourant.

Adverse   None known.
Effects

Typical   Bread
Products   Biscuits
                  Buns and cakes
                  Ice cream
                  Sweets
                  Vitamin and other
                      tablets

E171   **Titanium dioxide (C.I. 77891)**

Origin   Prepared from the naturally occurring mineral
                  ilmenite.

Function   White surface colour.

Adverse   None known.
Effects

Typical   Cottage cheese
Products   Vitamin tablets and
                      capsules
                  Horseradish cream
                  Horseradish sauce

E172   **Iron oxides, iron hydroxides (yellow/
                  brown: C.I. 77492; red: 77491;
                  brown: 77499)**

Origin   Naturally occurring pigments.

Function   Yellow, red, orange, brown and black colour.

Adverse   —
Effects

Typical   Salmon and shrimp
Products   paste
                  Packet dessert mix
                  Packet cake mix

E173   **Aluminium (C.I. 77000)**

Origin   Naturally occurring as the ore, bauxite.

Function   Metallic colour for surface only.

Adverse   Slight intestinal absorption of aluminium — any
Effects   absorbed is quickly excreted by healthy kidneys.

Typical   Solely for external
Products   covering of dragées
                  and decoration of
                  sugar-coated flour
                  confectionery, silvery
                  finish to pills and
                  tablets.

E174   **Silver (C.I. 77820)**

Origin   Naturally occurring metal.

Function   Metallic surface colour.

Adverse   Silver salts are toxic to bacteria and lower life-
Effects   forms. Long, regular consumption can lead to
                  argyria, a blue-grey skin, which is not dangerous.

Typical   Purely for dragées and
Products   sugar-coated flour
                  confectionery.

45

E175   **Gold (C.I. 77480)**

Origin   Naturally occurring metal.

Function   Metallic surface colour.

Adverse   Chemically very inactive, therefore harmless, but
Effects   expensive.

Typical   External application to
Products   dragées and sugar-
             coated flour
             confectionery

E180   **Pigment Rubine (Lithol Rubine BK; C.I. 15850)**

Origin   Synthetic, an azo dye (see glossary).

Function   Reddish colour.

Adverse   None known.
Effects

Typical   Solely for colouring the
Products   rind of cheese

Preservatives

E200	**Sorbic acid**
Origin	Occurs naturally in some fruits. May be obtained from the berries of mountain ash (*Sorbus aucuparia*), and can be manufactured synthetically. Products permitted to use E200 may also use the forms in E201, E202 and E203.
Function	Preservative, inhibiting the growth of yeasts and moulds.
Adverse Effects	Possible skin irritant.

Typical Products	Fermented milks Yogurt Gelatin capsules Fruit salads Sweets Soft drinks Bottled cheese spread Processed cheese slices Packet cake topping Surface of cheese Prepacked cake Frozen pizza

Candied peel
Canned cauliflower
Wine and cider
Dessert sauces
Fillings and toppings
Soup concentrates

Limits of Use Generally 1000mg/kg.

E201   Sodium sorbate

Origin   Sodium salt of sorbic acid which occurs naturally
or is manufactured synthetically.

Function   Preservative.

Adverse   None known.
Effects

Typical   Frozen pizza
Products

E202	**Potassium sorbate**

Origin Prepared from sorbic acid with potassium hydroxide.

Function Antifungal and antibacterial preservative, more soluble than sorbic acid.

Adverse Effects None known.

Typical Products
Fermented milk
Yogurt
Margarine/butter spread
Cheese spread
Salad dressing
Glacé cherries
Frozen vanilla pudding
Seafood dressing
Prepacked cakes
Tinned fruit pie fillings
Frozen pizza

E203	**Calcium sorbate**

Origin Synthetic.

Function Antifungal and antibacterial preservative.

Adverse Effects None known.

Typical Products
Fermented milk
    products
Yogurt

Typical Packet Soup (Vegetable)

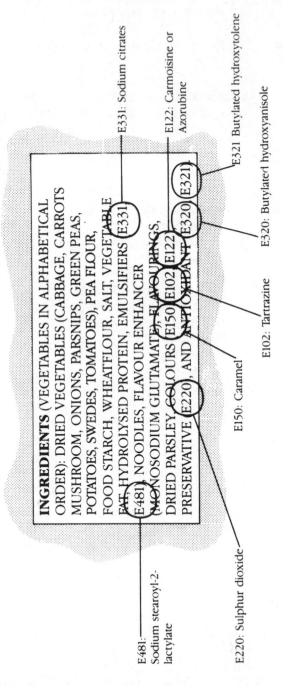

INGREDIENTS (VEGETABLES IN ALPHABETICAL ORDER): DRIED VEGETABLES (CABBAGE, CARROTS MUSHROOM, ONIONS, PARSNIPS, GREEN PEAS, POTATOES, SWEDES, TOMATOES), PEA FLOUR, FOOD STARCH, WHEATFLOUR, SALT, VEGETABLE FAT, HYDROLYSED PROTEIN, EMULSIFIERS (E331), E481, NOODLES, FLAVOUR ENHANCER MONOSODIUM GLUTAMATE, FLAVOURINGS, DRIED PARSLEY COLOURS (E150, E102, E122), AND ANTIOXIDANT (E220, AND ANTIOXIDANT (E320, E321).

E331: Sodium citrates

E122: Carmoisine or Azorubine

E321 Butylated hydroxytolene

E320: Butylated hydroxyanisole

E102: Tartrazine

E150: Caramel

E481: Sodium stearoyl-2-lactylate

E220: Sulphur dioxide

50

Typical Packet Pork Sausages

INGREDIENTS: PORK, WATER, RUSK, STARCH, TURKEY,
SALT, SPICES, SOYA PROTEIN CONCENTRATE,
SODIUM POLYPHOSPHATE, HERBS, DEXTROSE,
FLAVOUR ENHANCER (MONOSODIUM GLUTAMATE),
ANTIOXIDANT (E301) (E304) (E307), SUGAR, PRESERVATIVE
(E223), FLAVOURING, COLOUR (E128)

E223: Sodium
metabisulphite

E128: Red 2G

E301: Sodium-L-
ascorbate

E307: Synthetic
alpha-tocopherol

E304: 6-0-Palmitoyl-
L-ascorbic acid

Typical Packet Sliced Meat Loaf (Turkey and Ham)

INGREDIENTS: CURED TURKEY (TURKEY, SALT,
PRESERVATIVE: E250) HAM WITH PRESERVATIVE: E250),
PORK, WATER, STARCH, CASEINATES, SPICES, SODIUM
POLYPHOSPHATE, HERBS, ANTIOXIDANT: E301,
FLAVOUR ENHANCER: MONOSODIUM GLUTAMATE.

E250:
Sodium nitrite

E301: Sodium-L-
ascorbate

E250: Sodium
nitrite

E210   **Benzoic acid**

Origin   Occurs naturally in many edible berries but usually prepared synthetically.

Function   Preservatives — antibacterial and antifungal.

Adverse   People who suffer from asthma or who have
Effects   recurrent urticaria are likely to be sensitive to benzoic acid. It may also cause gastric irritation if consumed in large quantities. It has been reported to be responsible for neurological disorders and to react with the preservative sodium bisulphite (E222). See also pages 12-14.

Typical   Jams
Products   Beer
             Dessert sauces
             Flavouring syrups
             Fruit pulp and purée
             Fruit juice
             Marinated herring and
                 mackerel
             Pickles
             Salad cream and
                 dressing
             Fruit yogurt
             Coffee essence
             Products permitted to
                 use E210 may also
                 use E211, E212 and
                 E213.

Limits

These vary between 120mg/kg in fruit yogurt to 160mg/kg in soft drinks, 250mg/kg in products such as olives, sauces, salad cream and pickles to 450mg/kg in coffee essence to 850mg/kg in Indonesian sambal oelek right up to 2000mg/kg in unfermented grape juice intended for the sacrament.

E211   **Sodium benzoate**

Origin

The sodium salt of benzoic acid.

Function

Preservative — antibacterial and antifungal — effective only in slightly acid environment.

Adverse Effects

People who suffer from asthma, or who have recurrent urticaria, may be sensitive to Sodium benzoate and exhibit allergic reactions. See pages 12-14.

Typical Products

Caviar
Prawns
Sweets
Margarine
Fruit pies
Soft drinks
Oyster sauce
Salad dressing
Barbecue sauce
Mexican taco sauce
Cheesecake mix
Soya sauce
Orange squash

E212 **Potassium benzoate**

Origin The potassium salt of benzoic acid.

Function Preservative — antibacterial and antifungal.

Adverse Effects People who suffer from asthma or are allergic to aspirin or have recurrent urticaria, may be sensitive to potassium benzoate and show allergic reactions. See pages 12-14.

Typical Products —

E213 **Calcium benzoate**

Origin The calcium salt of benzoic acid.

Function Preservative — antibacterial and antifungal.

Adverse Effects People who suffer from asthma or recurrent urticaria, or are allergic to aspirin, may be sensitive to calcium benzoate and exhibit allergic reactions. See pages 12-14.

Typical Products —

E214 **Ethyl 4-hydroxybenzoate (Ethyl *para*-hydroxybenzoate)**

Origin Produced from benzoic acid.

Function Preservative — antibacterial and antifungal.

Adverse
Effects

Some people are hypersensitive to this substance, especially those sensitive to aspirin, asthmatics, those with recurrent urticaria. There may be skin sensitivity and/or a numbing effect on the mouth. See pages 12-14.

Typical
Products

Cooked prepacked
  beetroot
Chicory and coffee
  essence
Dessert sauces
Flavouring syrups
Flavourings
Liquid foam headings
Freeze drinks
Fruit based pie fillings
Preserved fruit — glacé,
  crystallized or
  drained
Fruit pulp or purée
Fruit juices
Marinated mackerel or
  herring
Salad cream
Jam

Products permitted to
  use E214 may also
  use E215.

Limits

160mg/kg in freeze drinks.
250mg/kg in beetroot, pickles and salad cream.
800mg/kg in flavouring syrups, fruit-based pie fillings, fruit pulp, juices, and concentrated soft drinks.
1000mg/kg in preserved fruit, marinated fish.
10000mg/kg in foam headings.

E215   **Ethyl 4-hydroxybenzoate, sodium salt (Sodium ethyl *para*-hydroxybenzoate)**

Origin   Produced from benzoic acid.

Function   Preservative — antibacterial and antifungal.

Adverse Effects   Some people are allergic to this substance, especially those sensitive to aspirin, asthmatics, those with recurrent urticaria. There may be skin sensitivity and/or a numbing effect on the mouth. See pages 12-14.

Typical Products   —

E216   **Propyl 4-hydroxybenzoate (n-Propyl p-hydroxybenzoate; Propyl *para*-hydroxybenzoate)**

Origin   Produced from benzoic acid.

Function   Preservative — antimicrobial.

Adverse Effects   Skin sensitivity with a numbing effect on the mouth. See pages 12-14.

Typical Products

Beer
Cooked prepacked
   beetroot
Coffee and chicory
   essence
Colouring matter in
   solution
Dessert sauces
Flavouring syrups
Freeze drinks
Fruit-based pie fillings
Crystallized, glacé or
   drained fruit
Fruit pulp or purée
Glucose and soft drinks
Marinated herring and
   mackerel
Pickles
Salad cream

Products permitted to
   use E216 may also
   use E217.

Limits

160mg/kg in freeze drinks.
250mg/kg in beetroot, pickles and salad cream.
800mg/kg in flavourings, fruit-based pie fillings,
   fruit pulp, juices and concentrated soft drinks.
1000mg/kg in preserved fruit and marinated fish.

E217   **Propyl 4-hydroxybenzoate, sodium salt (Sodium n-propyl p-hydroxybenzoate; Sodium propyl *para*-hydroxybenzoate)**

Origin   Produced from benzoic acid.

Function   Preservative — antimicrobial.

Adverse Effects   Allergic reactions to this substance may develop in asthmatics, those with recurrent urticaria or people sensitive to aspirin. There may be skin sensitivity and/or a numbing effect on the mouth. See pages 12-14.

Typical Products   —

E218   **Methyl 4-hydroxybenzoate (Methyl *para*-hydroxybenzoate)**

Origin   Synthetic.

Function   Preservative — antimicrobial agent.

Adverse Effects   Some people may exhibit allergic reactions to this substance mainly affecting the skin or mouth. See pages 12-14.

Typical
Products

Beer
Cooked prepacked
    beetroot
Coffee and chicory
    essence
Colouring matter in
    solution
Dessert sauces
Flavouring syrups
Freeze drinks
Fruit-based pie fillings
Crystallized, glacé or
    drained fruit
Fruit pulp or purée
Glucose and soft drinks
Marinated mackerel and
    herring
Pickles
Salad cream
Sauces
Snack meals
Soup concentrates

Products permitted to
    use E218 may also
    use E219.

Limits

160mg/kg in freeze drinks.
175mg/kg in snack meals and soup concentrates.
250mg/kg in beetroot, pickles and salad cream.
800mg/kg in flavourings, fruit-based pie fillings,
    fruit pulp, juices and concentrated soft drinks.
1000mg/kg in preserved fruit and marinated fish.

E219   **Methyl 4-hydroxybenzoate, sodium salt (Sodium methyl *para*-hydroxybenzoate; Sodium methyl hydroxybenzoate)**

Origin   Produced from benzoic acid.

Function   Preservative — active against fungi, and yeasts but less active against bacteria.

Adverse Effects   Allergic reactions have occurred when preparations containing hydroxybenzoates have been applied to the skin. Similar reactions have also occurred following intravenous or oral administration. Hydroxybenzoates have a numbing effect on the mouth. See pages 12-14.

Typical Products   —

E220   **Sulphur dioxide**

Origin   Occurs naturally but produced chemically by sulphur or gypsum combustion.

Function   Preservative; antioxidant; improving agent; bleaching agent (flour); vitamin C stabilizer.

Adverse Effects   Irritation of the alimentary food canal. Bleaching of flour has not been shown to be 100 per cent safe: it destroys much of the flour's vitamin E content. See also pages 12-14.

Typical
Products

Raspberry juice
Raspberry syrup
Fruit salads
Packet soup
Glacé cherries
Dried bananas and
    apricots
Blackcurrant jam
Desiccated coconut
Tinned cauliflower
Beer, wine, cider and
    cider vinegar
Candied peel
Tinned crabmeat
Fruit-based milk and
    cream desserts
Flavourings
Freeze drinks
Fruit-based pie fillings
Crystallized, glacé or
    drained fruit
Fruit pulp and purée
Fruit juices
Fruit spread
Powdered garlic
Gelatin
Dry root ginger
Glucose and soft drinks
Frozen mushrooms
Dehydrated vegetables
Sausage meat

Products permitted to
    contain E220 may
    also contain E221,
    E222, E223, E224
    and E227.

Limits   An astonishing range of additions are permitted, presumably calculated on the basis of the likely level of consumption, from 50mg/kg in desiccated coconut, 70mg/kg in beer through 100mg/kg in candied peel or canned cauliflower to 200mg/kg in cider, 2000mg/kg in most dried fruits right up to 30,000mg/kg in the dry enzyme from papaya, papain.

E221   **Sodium sulphite**

Origin   Synthetic.

Function   Preservative — antimicrobial; antioxidant in alkaline preparations. Main use is for 'fixing' photographs!

Adverse Effects   All sulphites may be dangerous to asthmatics.

Typical Products   Preserved egg yolk

E222   **Sodium hydrogen sulphite (Sodium bisulphite; acid sodium sulphite)**

Origin   Synthetic.

Function   Preservative; bleach.

Adverse Effects   A dose of 115mg per kilogram of body weight killed 50 per cent of a group of rats. All sulphites may be dangerous to asthmatics.

Typical   —
Products

E223   **Sodium metabisulphite
                (Diosodium pyrosulphite)**

Origin   Synthetic.

Function   Preservative; anti-oxidant.

Adverse   Ingestion of sodium metabisulphite and other
Effects   sulphites may cause gastric irritation due to
                liberation of sulphurous acid. Treatment of foods
                with sulphites reduces their thiamine (Vitamin B_1)
                content, so foods that contain a significant source
                of thiamine — meat, cereals, dairy products —
                should not be treated. Known cause of food
                aversion and allergic skin reactions. All sulphites
                may be dangerous to asthmatics.

Typical   Orange squash
Products   Pickled onions
                Pickled red cabbage
                Carton salad
                Packet mashed potatoes

E224 **Potassium metabisulphite (Potassium pyrosulphite)**

Origin Synthetic.

Function Preservative, especially in the Campden process for preserving fruit and home-made wine. Used to stop fermentation in breweries.

Adverse Effects Evidence in the United States indicates that asthmatics may be sulphite-sensitive. Some asthmatics have experienced wheezing, dyspnea (difficulty in breathing), cyanosis (deoxygenation of the blood), faintness and even unconsciousness after exposure to common sulphiting agents like potassium metabisulphite.

Typical Products Campden tablets

E226   **Calcium sulphite**

Origin   Synthetic.

Function   Preservative; firming agent.

Adverse
Effects   Ingestion of sulphite may cause gastric irritation due to the liberation of sulphurous acid. Treatment of foods with sulphite reduces their thiamine (vitamin B_1) content, so foods containing a significant source of thiamine (meat, cereals, dairy products) should not be treated. All sulphites may be dangerous to asthmatics.

Typical
Products   Cider

E227   **Calcium hydrogen sulphite**
            (Calcium bisulphite)

Origin   Synthetic.

Function   Preservative; prevents secondary fermentation; firming agent.

Adverse
Effects   Ingestion of sulphites may cause gastric irritation due to the liberation of sulphurous acid. Treatment of foods with sulphites reduces their thiamine (vitamin B_1) content, and may contribute to a vitamin deficiency. All sulphites may be dangerous to asthmatics.

Typical
Products   Beer

E230 **Biphenyl (Diphenyl)**

Origin Synthetic, produced by action of heat on benzene.

Function Preservative — antifungal.
Inhibits the growth of species of *Pencillium* which cause citrus fruits to decay. Can penetrate the skin of fruit and might be included in food or drink prepared from fruit.

Adverse Workers exposed to diphenyl reported nausea,
Effects vomiting and irritation to eyes and nose.

Typical Treatment of skins of
Products oranges, lemons,
grapefruit etc.
Can be partly removed
with detergent. Rinse
thoroughly afterwards.
Products permitted to
use E230 may also
use E231 and E232.

E231 **2-Hydroxybiphenyl (o-Phenyl phenol; Orthophenylphenol)**

Origin Synthetic, a substance used in the manufacture of rubber.

Function Preservative — antibacterial and antifungal.

Adverse Workers exposed to biphenyl (diphenyl) reported
Effects nausea, vomiting and irritation to eyes and nose.

Typical *Products*	Surfaces of citrus fruits and treatment of paper in which they are wrapped.

Products permitted to use E231 may also use E230 and E232.

E232

Sodium biphenyl-2-yl oxide (Sodium O-phenylphenol; Sodium orthophenylphenate)

Origin   Synthetic (with a strong smell of soap).

Function   Preservative — antifungal.

Adverse *Effects*   Alternative form of E231. Workers exposed to biphenyl (diphenyl) reported nausea, vomiting, and irritation to eyes and nose.

Typical *Products*   Penetration of the surface of citrus fruits may cause the substance to be present in marmalades and jams produced from the fruit.
Surface of citrus fruits and paper used to wrap citrus fruits.

E233   **2-(Thiazol-4-y1) benzimidazole (Thiabendazole)**

Origin   Synthetic.

Function   Preservative — fungicide.

Adverse Effects   None known.

Typical Products   Treatment of the skins of citrus fruits and impregnating the paper in which they are wrapped.

---

234   **Nisin**

Origin   A polypeptide antibiotic substance produced by the growth of a bacterium called *Streptococcus lactis*. Several strains of cheese starter organisms produce nisin.

Function   Preservative.

Adverse Effects   None known.

Typical Products   Cheese
Clotted cream
Cottage cheese
Canned foods

Under consideration by the E.E.C. for an 'E' prefix.

E236   Formic acid

Origin   Occurs naturally in the bodies of ants; produced synthetically.

Function   Preservative — antibacterial action.

Adverse
Effects   Very caustic to the skin and if absorbed has been known to cause urine disorders. Formic acid was formerly used as a diuretic.

Typical
Products   —

Not permitted in the
U.K.

E237   Sodium formate

Origin   The sodium salt manufactured from formic acid.

Function   Preservative.

Adverse
Effects   Has diuretic properties and was formerly used for this purpose.

Typical
Products   —

Not permitted in the
U.K.

E238   Calcium formate

Origin   The calcium salt of formic acid.

Function   Preservative.

Adverse   Has diuretic properties and was formerly used
Effects   for this purpose.

Typical   —
Products
                Not permitted in the
                  U.K.

E239   Hexamine (Hexamethylenetetramine)

Origin   Synthetic, a derivative of benzene.

Function   Preservative — fungicide.

Adverse   Gastro-intestinal upsets may result from
Effects   prolonged use of hexamine by the production
                of formaldehyde. In addition, the urinary system
                may be affected and less frequently, skin rashes
                may occur. In experiments with animals,
                hexamine caused gene mutation and is suspected
                of being carcinogenic.

Typical   Marinated herrings and
Products   mackerel
                Provolone cheese

Limits   25mg/kg (expressed as formaldehyde)

E249   **Potassium nitrite**

Origin   Naturally occurring.

Function   Food preservative, curing agent.

Adverse   Should not be used in food for babies under six
Effects   months. Destroys red blood corpuscles. Reacts
                with amines to form nitrosamines which have
                been shown to be potentially carcinogenic.
                Preservatives in sausages have been linked with
                asthma.

Typical   Cooked meats
Products   Sausages

                Sodium nitrite (E250)
                    may be used instead.

E250   **Sodium nitrite**

Origin   Not naturally occurring; derived from sodium
                    nitrate by chemical or bacterial action.

Function   Food preservative (inhibiting the growth of
                    Clostridium botulinum the bacterium responsible
                    for botulism); curing salt.

Adverse   Nitrites may cause nausea and vomiting, dizziness,
Effects   headaches, de-oxygenation of the blood, low
                    blood pressure and collapse of the circulatory
                    system. Nitrites form minute traces of
                    nitrosamines in the stomach and these have been
                    shown to be carcinogenic in animals but not yet
                    in man. They are prohibited for sale in foods for
                    babies and young children. Preservatives in
                    sausages were shown to be responsible for
                    asthma in a 57-year-old woman. See also pages
                    12-14.

Typical   Cured meat and cured
Products   meat products
                    Salted meat to fix the
                        red colour
                    Pork sausage
                    Packet bacon steaks
                    Turkey and ham loaf
                    Smoked frankfurters
                    Bacon
                    Ham
                    Tongue
                    Pressed meat
                    Tinned meat
                    Frozen pizza

Limits (E250   50mg/kg in cheese other than Cheddar, Cheshire
and E251)   or soft cheeses (of which only 5mg may be E250)
150mg/kg in cured meat (of which not more than 50mg may be E250)
400mg/kg in salami and similar acidified or fermented meat products (of which not more than 50mg may be E250)
500mg/kg in uncooked bacon and ham and cooked unsealed bacon and ham (of which not more than 200mg may be E250).

E251	**Sodium nitrate (Chile saltpetre)**

Origin   Naturally occurring mineral (especially in the Atacama desert, Chile).

Function   Preservative; curing salt.

Adverse Effects   Nitrates are capable of being converted to nitrites either when food spoils or by bacteria in the stomach (especially in tiny babies). Nitrites can cause deoxygenation of the blood or form minute amounts of nitrosamines which may be carcinogenic. See also pages 12-14.

Typical Products   Bacon
Pressed meats
Ham
Tongue
Beef
Canned meat
Cheese
Frozen pizza

Potassium nitrate (E252)
   may be used instead.

E252   **Potassium nitrate (Saltpetre)**

Origin   Naturally occurring mineral, or artificially manufactured from waste animal and vegetable material.

Function   Food preservative, curing salt, one of the oldest and most effective ways of preserving meats.

Adverse   Gastro-enteritis with severe abdominal pain,
Effects   vomiting, vertigo, muscular weakness, and irregular pulse can occur. Potassium nitrate may be reduced to potassium nitrite in the gut by bacterial action and this once absorbed can affect the haemoglobin in the red blood corpuscles preventing it carrying oxygen. Nitrites can produce minute amounts of nitrosamines which may be carcinogenic in man. Nevertheless without the nitrates and nitrites there would be many deaths from the growth of toxic micro-organisms in meats.

Typical   Cured meats
Products   Sausages
              Smoked frankfurters
              Bacon, ham, tongue
              Pressed meats
              Tinned meats

E260   Acetic acid

Origin   Manufactured by the destructive distillation of wood. The acetic acid in vinegar is formed by the action of the bacterium *Acetobacter* on the alcohol in beer for malt vinegar or cider or wine for those vinegars.

Function   Antibacterial and at 5 per cent concentration may be bactericidal; substance permitted to stabilize the acidity of food; diluent for colouring matter.

Adverse   No toxicological problems are known.
Effects

Typical   Foods which may
Products   provide a suitable
                environment for
                certain bacteria
            Pickles
            Chutneys
            Cheese
            Salad cream
            Fruit sauce
            Brown sauce
            Spicy brown sauce
            Mint sauce and jelly
            Horseradish cream
            Dilute acetic acid at
                about 5 per cent is
                called 'non-brewed
                condiment'

Typical Powdered Orange Drink

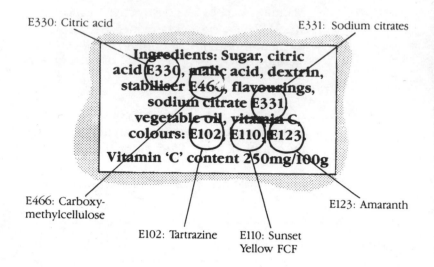

E330: Citric acid

E331: Sodium citrates

Ingredients: Sugar, citric acid (E330), malic acid, dextrin, stabiliser E466, flavourings, sodium citrate (E331) vegetable oil, vitamin C, colours: (E102) (E110) (E123).

Vitamin 'C' content 250mg/100g

E466: Carboxy-
methylcellulose

E123: Amaranth

E102: Tartrazine

E110: Sunset
Yellow FCF

Typical Cartonned Orange Drink

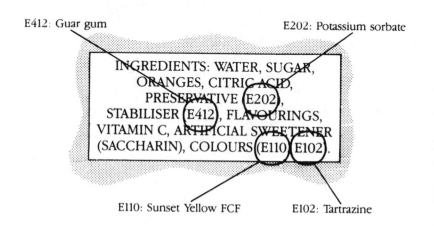

E412: Guar gum

E202: Potassium sorbate

INGREDIENTS: WATER, SUGAR, ORANGES, CITRIC ACID, PRESERVATIVE (E202), STABILISER (E412), FLAVOURINGS, VITAMIN C, ARTIFICIAL SWEETENER (SACCHARIN), COLOURS (E110) (E102).

E110: Sunset Yellow FCF

E102: Tartrazine

Typical Frothy Milk Shake (Strawberry)

E412 Guar gum

E1422:
see pages
18-19

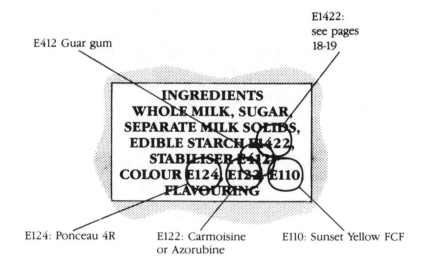

**INGREDIENTS
WHOLE MILK, SUGAR,
SEPARATE MILK SOLIDS,
EDIBLE STARCH E1422,
STABILISER E412,
COLOUR E124, E122, E110
FLAVOURING**

E124: Ponceau 4R

E122: Carmoisine
or Azorubine

E110: Sunset Yellow FCF

E261   Potassium acetate

Origin   The potassium salt of acetic acid, E260.

Function   To preserve natural colour of plant and animal tissues; buffer.

Adverse Effects   Potassium salts, taken by mouth in healthy people cause little toxicity since potassium is rapidly excreted in the urine, but should be avoided by people with impaired kidneys.

Typical Products   —

E262   Sodium hydrogen diacetate (Sodium diacetate)

Origin   Synthetic.

Function   Preservative — antimicrobial inhibitor especially against the spores of *Bacillus mesentericus* and *B. subtilis*. These spores are heat-resistant, and if present in bread and permitted to germinate, convert the bread into sticky yellow patches, which are capable of being pulled into long threads, hence the term 'rope-forming' micro-organisms.

Adverse Effects   None known.

Typical Products   Bread
Shaped crisps

262   Sodium acetate (anhydrous) and Sodium acetate

Origin   The sodium salt of acetic acid, E260.

Function   Buffer (acid or alkaline stabilizer).

Adverse Effects   Too much sodium in the body could be dangerous in babies and young children by increasing the body fluid salt concentrations either within or between the cells.

Typical Products   —

Under consideration by the E.E.C. for an 'E' prefix.

E263   Calcium acetate

Origin   The calcium salt of acetic acid, E260.

Function   Food preservative, preventing mould growth; sequestrant; firming agent.

Adverse Effects   None known.

Typical Products   Packet cheesecake mix
Quick setting jelly mix

E270   Lactic acid

Origin
Naturally occurring substance, produced by milk-souring organisms, also manufactured by fermentation of a carbohydrate with *Bacillus acidilactic* at high temperatures.

Function
Food preservative; capable of increasing the antioxidant effect of other substances; acid and flavouring.

Adverse Effects
Could cause problems in very young or premature babies who may have difficulty metabolizing it. No toxicological problems with adults.

Typical Products
Soft margarine
Carbonated drinks
Infant milks
Confectionery
Carton salad in dressing
Salad dressing
Pickled red cabbage
Bottled cheese spread
Sauce tartare

E280   **Propionic acid**

Origin   A naturally occurring fatty acid, one of the
                products of digestion of cellulose by the gut-
                inhabiting bacteria of herbivorous animals.
                Produced by fermentation for commercial use.

Function   Food preservative — antifungal agent against
                three families of fungi.

Adverse   No known toxicological problems.
Effects

Typical   Baking and dairy
Products   products
                Pizza
                Christmas puddings

                Substances permitted to
                   use E280 may also
                   use E281, E282 and
                   E283.

Limit   300mg/kg (calculated on the weight of flour)

E281   **Sodium propionate**

Origin   The sodium salt of propionic acid.

Function   Food preservative — an antifungal agent against
                  three families of moulds.

Adverse   One report suggests that sodium propionate is
Effects   involved with certain migraine headaches.

Typical   Dairy and bakery
Products   products

E282   **Calcium propionate**

Origin   Occurs naturally in Swiss cheese; prepared
                  commercially from propionic acid.

Function   Preservative — antimicrobial mould inhibitor,
                  especially 'rope' micro-organisms, which occur
                  in bread. The spores of *Bacillus mesentericus* and
                  B. subtilis are heat resistant and if present in bread
                  and permitted to germinate convert the bread into
                  sticky yellow patches, which are capable of being
                  pulled into long threads.

Adverse   No known toxicological problems.
Effects

Typical   Dairy and baking
Products   products
                  Frozen pizza

E283   **Potassium propionate**

Origin   The potassium salt of propionic acid.

Function   Preservative — mould inhibitor.

Adverse Effects   None known.

Typical Products   Dairy and bakery products
Christmas puddings

E290   **Carbon dioxide**

Origin   Natural gas, present in atmospheric air but produced by fermentation or the action of acid on a carbonate or as a by-product in the manufacture of lime.

Function   Preservative; coolant; freezant (liquid form); packaging gas.

Adverse Effects   Some carbonates in the stomach increase the secretion of gastric acid and promote absorption of liquid by the mucous membranes, increasing the effect of alcohol.

Typical Products   Fizzy and effervescent drinks

296   Malic acid (DL-or-L-)

Origin   L form present in apples, pears and many other
               temperate fruits. DL form produced chemically.

Function   Acid, flavouring.

Adverse   None known.
Effects

Typical   Tinned oxtail soup
Products   Shaped crisps
               Low calorie orange
                   squash
               Packet spaghetti sauce
                   mix

               Under consideration by
                   the E.E.C. for an 'E'
                   prefix.

---

297   Fumaric acid

Origin A naturally occurring organic acid especially important in cell respiration. Prepared by fermentation for commercial use.

Function Acidifier and flavouring agent; raising agent and antioxidant in baked goods.

Adverse Effects None known.

Typical Products Packet cheesecake mix
Yogurt whip

Under consideration by
the E.E.C. for an 'E'
prefix.

Permitted Anti-Oxidants

E300   **L-Ascorbic acid (Vitamin C)**

Origin   Naturally occurring substance in many fresh fruits and vegetables; also manufactured by biological synthesis.

Function   Vitamin C; antioxidant in emulsions of fats and oils and in iron mixtures; browning inhibitor in unprocessed cut fruits, fruit pulp and juices; improving agent for flour; meat colour preservative.

Adverse Effects   Usually well tolerated. Large doses may cause diarrhoea and/or dental erosion. More than 10g per day could result in kidney stones in susceptible people.

Typical
Products

Concentrated fruit
   drinks
Butter
Beer
Soft and fizzy drinks
Frozen egg products
Powdered and
   concentrated milk
Fruit jams and
   preserves
Frozen croquette
   potatoes
Dried potatoes (where
   its presence is
   desirable as a
   replacement for the
   vitamin C lost in
   processing.)

E301

Sodium L-ascorbate (Vitamin C; Sodium L-(+)-ascorbate)

Origin

Prepared synthetically, the sodium salt of ascorbic acid.

Function

Vitamin C; antioxidant; colour preservative.

Adverse Effects

No toxicological problems in standard doses.

Typical Products

Scotch eggs
Sausages
Turkey and ham loaf
Smoked frankfurters

87

E302   Calcium L-ascorbate (Calcium ascorbate)

Origin   Prepared synthetically.

Function   Vitamin C; antioxidant; meat colour preservative.

Adverse Effects   None known.

Typical Products   Scotch eggs

E304   6-0-Palmitoyl-L-ascorbic acid (Ascorbyl palmitate)

Origin   Synthetic.

Function   Antioxidant; colour preservative.

Adverse Effects   None known.

Typical Products   Scotch eggs
Sausages
Chicken stock tablets

E306   Extracts of natural origin rich in tocopherols (Vitamin E)

Origin   Extract of soya bean oil, wheat germ, rice germ, cottonseed, maize and green leaves, distilled in a vacuum.

Function   Vitamin; antioxidant.

Adverse Effects   None known in food use.

Typical Products   Packet dessert topping
Vegetable oils

E307   Synthetic *alpha*-tocopherol (Vitamin E; DL-*alpha*-tocopherol)

Origin   Synthetic.

Function   Antioxidant; vitamin.

Adverse Effects   None known in food use.

Typical Products   Sausages

E308   **Synthetic *gamma*-tocopherol (Vitamin E; DL-*gamma*-tocopherol)**

Origin   Synthetic.

Function   Antioxidant; vitamin.

Adverse Effects   None known in food use.

Typical Products   —

E309   **Synthetic *delta*-tocopherol (Vitamin E; DL-*delta*-tocopherol)**

Origin   Synthetic.

Function   Antioxidant; vitamin.

Adverse Effects   None known in food use.

Typical Products   —

E310 **Propyl gallate (Propyl 3,4,5, trihyroxybenzoate)**

Origin Prepared synthetically.

Function Antioxidant in oils and fats.

Adverse Effects All alkyl gallates may cause gastric irritation and problems for people who suffer from asthma or are sensitive to aspirin. One report implies that this substance could cause reproductive failures and liver damage. Not permitted in foods intended for babies or young children. See also pages 12-14.

Typical Products Vegetable oils and
    shortenings
Dry breakfast cereals
Margarine
Instant potatoes
Snack foods
Chewing gum

E311   **Octyl gallate**

Origin   Prepared synthetically.

Function   Antioxidant.

Adverse   All alkyl gallates may cause gastric irritation and
Effects   problems for people who suffer from asthma or
                  are sensitive to aspirin. Not permitted in foods
                  intended for babies or young children. See also
                  pages 12-14.

Typical   —
Products

E312   **Dodecyl gallate (Dodecyl 3,4,5, -
                  trihydroxybenzoate)**

Origin   Synthetically prepared.

Function   Antioxidant.

Adverse   All alkyl gallates may cause gastric irritation and
Effects   problems for people who suffer from asthma or
                  are sensitive to aspirin. Not permitted in foods
                  intended for babies or young children. See also
                  pages 12-14.

Typical   —
Products

E320   **Butylated hydroxyanisole (BHA)**

Origin   Prepared synthetically.

Function   Antioxidant for oils and fats either alone or with
                a gallate and a synergist (see glossary) e.g., citric
                acid or phosphoric acid. Delays or retards or
                prevents the development in food of rancidity
                or other flavour deterioration due to oxidation.
                It is heat resistant, so effective in baked products.

Adverse   Raises the lipid and cholesterol levels in the blood.
Effects   Can induce the formation of metabolizing
                enzymes in the liver with an increased risk of
                breakdown of important substances in the body
                such as vitamin D. Not permitted in foods
                intended for babies or young children except to
                preserve added vitamin A. See also pages 12-14.
                    American work on animals suggests that E320
                and E321 could possibly protect against some
                carcinogens.

Typical   Biscuits
Products   Sweets
                Raisins
                Fruit pies
                Soft drinks
                Margarine
                Cheese spread
                Sachet marinade
                Beef stock cubes
                Savoury rice
                Butter
                Packet convenience
                    foods

E321   **Butylated hydroxytoluene (BHT)**

Origin   Prepared synthetically.

Function   Antioxidant for food oils and fats.

Adverse   Some people are sensitive to the presence of BHT
Effects   in foods, developing rashes. It can cause the
                  development of metabolizing enzymes in the liver
                  which can increase the rate of breakdown of
                  other substances in the body, e.g. vitamin D.
                  Various reports have linked this additive with
                  possible reproductive failures, behavioural effects
                  and blood cell changes. Not permitted in foods
                  intended for babies or young children except to
                  preserve added vitamin A. See also pages 12-14.
                      American research suggests that BHT increases
                  the life-span of experimental mice. A Danish
                  report, however, claims BHT increases tumour
                  formation in rats. As some manufacturers use, and
                  others do not use, BHT for similar products, the
                  commercial necessity for BHT has been
                  questioned.

Typical   Sachet marinade
Products   Packet cake mix
                  Savoury rice
                  Soft and other margarines,
                     shortenings and vegetable
                     oils
                  Crisps
                  Salted peanuts
                  Potato rings
                  Gravy granules
                  Dehydrated mashed potato
                  Dry breakfast cereals
                  Chewing gum
                  Packet convenience foods

Emulsifiers, Stabilizers and Others

E322   **Lecithins**

Origin   Most commercial lecithin is obtained from soya beans. Other commercial sources are egg yolk, leguminous seeds including peanuts and maize, although lecithin is present in all living cells.

Function   Surfactant (surface-active agent); emulsifier; stabilizer; antioxidant and viscosity reducer in chocolate; plasticizer (additional thickness for fats).

Adverse   No known toxicological problems. Used
Effects   experimentally to treat senile dementia and to mobilize fats in the body.

Typical   Chocolate
Products   Powdered milks
                  Soft margarine
                  Confectionery
                  Dessert mixtures
                  Packet trifle mix
                  Vermicelli
                  Chocolate cake
                      covering
                  Yogurt whip
                  Chocolate biscuits
                  Lecithin granules

E325 **Sodium lactate**

Origin The sodium salt of lactic acid.

Function Humectant and substitute for glycerol; synergistic effect on other substances by increasing antioxidant effect.

Adverse Effects Could have a certain toxicity for very young children. No toxicological problems known with adults.

Typical Products Confectionery
Cheese

E326 **Potassium lactate**

Origin The potassium salt of lactic acid.

Function Capable of increasing the antioxidant effect of other substances; buffer.

Adverse Effects None known.

Typical Products —

E327 **Calcium lactate**

Origin The calcium salt of lactic acid.

Function Antioxidant; capable of increasing antioxidant effect of other substances; buffer; firming agent.

Adverse Effects Not known.

Typical Products Packet lemon meringue pie mix

E330   **Citric acid**

Origin   Occurs naturally in lemon and other citrus juices; prepared commercially by the fermentation of molasses with strains of *Aspergillus niger*.

Function   As a synergist to enhance the effectiveness of antioxidants: prevents discolouration of fruit, development of 'off' flavours and retains vitamin C. Stabilizes the acidity of food substances; sequestrant; flavouring; helps jam to set.

Adverse   Citric acid taken in very large quantities may
Effects   occasionally cause erosion of the teeth and have a local irritant action.

Typical   Biscuits
Products   Tinned vegetables
                  Tinned fruit
                  Non-alcoholic drinks
                  Frozen croquette
                      potatoes
                  Frozen potato waffles
                  Tinned sauces
                  Treatment of raisins
                  Ice cream
                  Packet cake mix
                  Packet soup mix
                  Sorbet mix
                  Wine and cider
                  Flavouring in drinks
                      and confectionery
                  Jams and jelly preserves
                  Frozen fish, especially
                      herrings, shrimps
                      and crab
                  Bakery products
                  Cheese

Pasteurized processed
  cheese
Cheese spread
Cream cottage cheese

E331   **Sodium citrates**
E331(a)   **Sodium dihydrogen citrate**
            (*mono*Sodium citrate)

Origin   A sodium salt of citric acid.

Function   Synergistic effect on other antioxidants; buffer; emulsifying salt; sequestrant; added to infant milk feeds and invalid food to prevent formation of large curds.

Adverse   None known.
Effects

Typical   Ice cream
Products   Sweets
            Packet Black Forest
               gateau mix

E331(b) *di*Sodium citrate

Origin   A sodium salt of citric acid.

Function   Antioxidant; synergistic effect on other
               antioxidants; buffer; emulsifying salt.

Adverse   None known.
Effects

Typical   Wines
Products   Fizzy drinks
               Processed cheese slices

E331(c) *tri*Sodium citrate (Citrosodine)

Origin   A sodium salt of citric acid.

Function   Antioxidant; buffer; emulsifying salt; sequestrant.

Adverse   None known.
Effects

Typical   Processed cheese
Products

E332   **Potassium dihydrogen citrate**
(mono Potassium citrate)

Origin   A potassium salt of citric acid.

Function   Buffer; emulsifying salt.

Adverse   None known; potassium is rapidly excreted in the
Effects   urine in healthy individuals.

Typical   Sterilized and UHT
Products   cream
Condensed milk
Dried milk
Cheese
Reduced-sugar jam

E332   ***tri*Potassium citrate**
(Potassium citrate)

Origin   A potassium salt of citric acid.

Function   Antioxidant; buffer; emulsifying salt; sequestrant.

Adverse   None in foods; in therapeutic amounts may make
Effects   the skin sensitive and cause mouth ulcers to
develop.

Typical   Confectionery
Products   Wines
Fizzy drinks
Cheeses
Rum sauce
Crisps
Biscuits
Packet convenience
   dessert mixes and
   toppings

E333   *mono*, *di*, and *tri* **Calcium citrate**

Origin   Calcium salts of citric acid.

Function   Buffers; firming agents; emulsifying salts.

Adverse Effects   None in foods; in therapeutic amounts may induce the formation of mouth ulcers.

Typical Products
Wines
Fizzy drinks
Confectionery
Cheeses

E334   **L-(+)-Tartaric acid**

Origin   Occurs naturally in grapes; sometimes deposited as crystals in wine. Manufactured as a by-product of the wine industry.

Function   Antioxidant; capable of increasing the anxioxidant effect of other substances (synergist); acid; sequestrant (see glossary); diluent for food colours.

Adverse Effects   No known toxicological risks. Strong solutions of tartaric acid are mildly irritant and if ingested undiluted may cause gastro-enteritis.

Typical Products
Confectionery
Jams
Jellies
Marmalades
Fizzy drinks

E335   ***mono*Sodium L-(+)-tartrate
and *di*Sodium L-(+)-tartrate**

Origin   Manufactured from tartaric acid.

Function   Antioxidant and capable of increasing the antioxidant effect of other substances (synergist); buffer; emulsifying salt; sequestrant.

Adverse   No known toxicological risks.
Effects

Typical   Confectionery
Products   Jams
Jellies
Marmalades
Fizzy drinks

Typical Fruit Pie (Apricot)

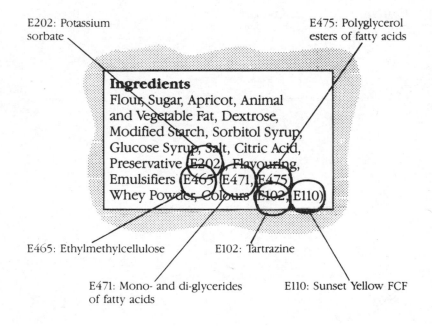

E202: Potassium
sorbate

E475: Polyglycerol
esters of fatty acids

Ingredients
Flour, Sugar, Apricot, Animal
and Vegetable Fat, Dextrose,
Modified Starch, Sorbitol Syrup,
Glucose Syrup, Salt, Citric Acid,
Preservative (E202), Flavouring,
Emulsifiers (E465, E471, E475),
Whey Powder, Colours (E102, E110)

E465: Ethylmethylcellulose

E102: Tartrazine

E471: Mono- and di-glycerides
of fatty acids

E110: Sunset Yellow FCF

Typical Dessert Mix (Strawberry)

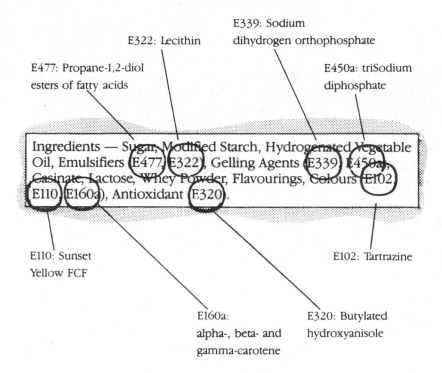

E322: Lecithin

E339: Sodium
dihydrogen orthophosphate

E477: Propane-1,2-diol
esters of fatty acids

E450a: triSodium
diphosphate

Ingredients — Sugar, Modified Starch, Hydrogenated Vegetable Oil, Emulsifiers (E477, E322), Gelling Agents (E339, E450a), Casinate, Lactose, Whey Powder, Flavourings, Colours (E102, E110, E160a), Antioxidant (E320).

E110: Sunset
Yellow FCF

E102: Tartrazine

E160a:
alpha-, beta- and
gamma-carotene

E320: Butylated
hydroxyanisole

E336

***mono* Potassium L-(+)-tartrate
(Potassium hydrogen tartrate;
Cream of tartar; Potassium
acid tartrate)**

Origin

Prepared from tartaric acid.

Function

Acid; buffer; emulsifying salt; raising agent for flour, often used with 500 (Sodium bicarbonate) because it works slowly and gives a more prolonged evolution of carbon dixoide; inverting agent for sugar in boiled sweet manufacture.

Adverse Effects

None known; potassium salts are readily excreted by healthy kidneys. The only people at risk are those whose kidney or liver functions are impaired.

Typical Products

Packet lemon meringue
   pie mix
Packet lemon meringue
   crunch mix

E336   *di*Potassium L-(+)-tartrate

Origin   Manufactured from tartaric acid.

Function   Antioxidant and capable of increasing the antioxidant effect of other substances (synergist); buffer; emulsifying salt.

Adverse Effects   No known toxicological risks.

Typical Products
Jelly part of packet
    trifle mix
Packet meringue
    crunch mix

E337   Potassium sodium L-(+)-tartrate (Sodium and potassium tartrate; Sodium potassium tartrate: Rochelle salt)

Origin   The potassium and sodium salts of tartaric acid.

Function   Buffer; emulsifying salt; stabilizer; capable of increasing antioxidant effect of other substances (synergist).

Adverse Effects   No known toxicological problems. The tartrates of the alkali metals are less readily absorbed than the citrates.

Typical Products
Meat and cheese
    products

107

E338   **Orthophosphoric acid
(Phosphoric acid)**

Origin   Prepared by the action of sulphuric acid on
tricalcium phosphate from phosphate rock
deposits in the United States.

Function   To speed penetration of brine; capable of
increasing the antioxidant effect of other
substances (synergist); acid; flavouring and
sequestrant.

Adverse   None known.
Effects

Typical   Fizzy drinks
Products   Cooked meats and
                  sausages
              Ham
              Cheese

E339(a) **Sodium dihydrogen orthophosphate N.B. E339, (a), (b), (c) are all classified as E339**

Origin Prepared from phosphoric acid.

Function To improve texture and prevent seepage of serum from foods; to speed penetration of brine; antioxidant synergist; buffer.

Adverse Effects None known.

Typical Products
Cooked meats
Sausages
Fizzy drinks
Ham
Packet cheesecake mix
Packet lemon pie filling

E339(b) **diSodium hydrogen orthophosphate**

Origin Prepared from phosphoric acid.

Function Buffer; nutrient; gelling agent; stabilizer.

Adverse Effects None known.

Typical Products
Butter/margarine
    compound
Cooked meats
Ham
Sausages

E339(c)   *tri*Sodium orthophosphate (Sodium phosphate tribasic)

Origin   Prepared from phosphoric acid.

Function   Clarifying sugar; emulsifying salt; anti-caking agent; buffer; antioxidant synergist.

Adverse Effects   None known.

Typical Products
Processed cheese
Cooked meats
Ham
Sausages
Cheese spread
Processed cheese slices
Bottled cheese spread
Instant whipped
    desserts

E340(a)   **Potassium dihydrogen orthophosphate (Potassium phosphate monobasic) N.B. E340 (a), (b), (c) are all classified as E340**

Origin   Prepared from phosphoric acid.

Function   Buffer; sequestrant; emulsifying salt; antioxidant synergist.

Adverse Effects   None known.

Typical Products	Jelly part of packet trifle mix Dessert topping

E340(b)   **diPotassium hydrogen orthophosphate (Potassium phosphate dibasic)**

Origin   Prepared from phosphoric acid.

Function   Buffer; emulsifying salt; antioxidant synergist.

Adverse Effects   None known.

Typical Products   Non-dairy powdered coffee creamers

E340(c)   **triPotassium orthophosphate (Potassium phosphate tribasic)**

Origin   Prepared from phosphoric acid.

Function   Emulsifying salt; antioxidant synergist; buffer.

Adverse Effects   None known.

Typical Products   —

E341(a) **Calcium tetrahydrogen**
diorthophosphate (acid Calcium
phosphate; ACP)
N.B. E341 (a), (b), (c) are all classified
as E341

Origin Calcium phosphate (apatite) occurs naturally and
derivitives of this are produced chemically.

Function Improving agent; buffer; firming agent;
emulsifying salt; sequestrant; yeast food; raising
agent in nearly all baking powders; antioxidant
synergist.

Adverse None known.
Effects

Typical Short pastry mix
Products Baking powder

E341(b) **Calcium hydrogen orthophosphate**
(Calcium phosphate dibasic)

Origin Prepared chemically from naturally derived
calcium phosphate.

Function Emulsifying salt; firming agent; yeast food;
nutrient mineral supplement in cereals and other
foods; antioxidant synergist; animal feed
supplement; abrasive in toothpaste and reduces
the incidence of dental caries when added to
sweets.

Adverse None known.
Effects

Typical Tinned cherry pie
Products filling

E341(c)   *tri*Calcium diorthophosphate

Origin   Prepared chemically from naturally derived calcium phosphate.

Function   Anti-caking agent; buffer; antioxidant synergist; emulsifying salt; nutrient yeast food; diluent for vegetable extracts; abrasive in toothpastes; clarifying sugar syrups.

Adverse   None known.
Effects

Typical   Cake mixes
Products

350   **Sodium malate**

Origin   A sodium salt of malic acid.

Function   Buffer.

Adverse   None known.
Effects

Typical   —
Products
                Under consideration by
                the E.E.C. for an 'E'
                prefix.

350   Sodium hydrogen malate

Origin   A sodium salt of malic acid.

Function   Buffer.

Adverse Effects   None known.

Typical Products   —

Under consideration by
the E.E.C. for an 'E'
prefix.

351   Potassium malate

Origin   The potassium salt of malic acid.

Function   Buffer.

Adverse Effects   None known.

Typical Products   —

Under consideration by
the E.E.C. for an 'E'
prefix.

352 Calcium malate

Origin   A calcium salt of malic acid.

Function   Buffer; firming agent.

Adverse Effects   None known.

Typical Products   —

Under consideration by the E.E.C. for an 'E' prefix.

---

352 Calcium hydrogen malate

Origin   A calcium salt of malic acid.

Function   Firming agent.

Adverse Effects   None known.

Typical Products   —

Under consideration by the E.E.C. for an 'E' prefix.

---

353   Metatartaric acid

Origin   Prepared from tartaric acid.

Function   Sequestrant.

Adverse   None known.
Effects

Typical   Wine
Products
                Under consideration by
                    the E.E.C. for an 'E'
                    prefix.

---

355   Adipic acid (Hexanedioic acid)

Origin   An organic acid which occurs in many living cells
                and especially in beet juice. Prepared synthetically
                for commercial use by oxidizing cyclohexanol
                with nitric acid..

Function   Acidulating agent; flavouring agent; raising agent
                in baking powders, since unlike tartaric acid
                (E334), cream of tartar (E336) and phosphates,
                adipic acid is not hydroscopic (water attracting).

Adverse   None known.
Effects

Typical   —
Products
                Under consideration by
                    the E.E.C. for an 'E'
                    prefix.

---

363 Succinic acid

Origin	Occurs naturally in fossils, fungi and lichens but prepared for commercial use from acetic acid.
Function	Acid.
Adverse Effects	None known.

Typical Products	—
	Under consideration by the E.E.C. for an 'E' prefix.

370 1,4-Heptonolactone

Origin	Prepared synthetically.
Function	Acid; sequestrant.
Adverse Effects	None known.

Typical Products	—
	Under consideration by the E.E.C. for an 'E' prefix.

375   **Nicotinic acid (Niacin; Nicotinamide)**

Origin   Occurs naturally in yeast, liver, rice polishings and lean meats, although it is prepared for commercial use by the oxidation of nicotine with concentrated nitric acid.

Function   B Vitamin; colour protector.

Adverse Effects   Nicotinic acid can dilate the blood vessels and if given in therapeutic doses it may produce flushing of the face, and pounding in the head and a sensation of heat. It is perfectly safe in normal use.

Typical Products
Bread
Flour
Breakfast cereals

Under consideration by the E.E.C. for an 'E' prefix.

380   ***triAmmonium* citrate**

Origin   Prepared from citric acid.

Function   Buffer; emulsifying salt.

Adverse Effects   None known.

Typical Products   —

Under consideration by the E.E.C. for an 'E' prefix.

381 Ammonium ferric citrate (Ferric ammonium citrate)

Origin   Prepared from citric acid.

Function   Dietary iron supplement; used medically for raising the level of red blood cells.

Adverse Effects   None known.

Typical Products   Iron tablets
Infant milk formulae
Bread flour (not 100% wholemeal)

Under consideration by
the E.E.C. for an 'E'
prefix.

---

381 Ammonium ferric citrate, green

Origin   Prepared from citric acid.

Function   Dietary iron supplement.

Adverse Effects   None known.

Typical Products   —

Under consideration by
the E.E.C. for an 'E'
prefix.

119

385   **Calcium disodium ethylenediamine —
NNN'N' tetra-acetate (Calcium
disodium EDTA)**

Origin   Prepared synthetically.

Function   Sequestrant; chelating substance (binds to free
metals promoting stabilization of products).

Adverse   In larger quantities than would be the case as a
Effects   food additive calcium disodium EDTA has caused
vomiting, diarrhoea and abdominal cramps.
There has been some suggestion that the additive
may interfere with the body's absorption of
essential trace elements, such as iron, zinc and
copper.

Typical   Salad dressings
Products   Alcoholic beverages

Under consideration by
the E.E.C. for an 'E'
prefix.

E400 **Alginic acid**

Origin Extracted from brown seaweeds, mainly *Laminaria*, growing off the west coasts of Scotland and Ireland.

Function Emulsifier; stabilizer; gelling agent; thickener.

Adverse Natural product which produces no known
Effects toxicological risks.

Typical Ice cream (as stabilizing
Products   colloid) ensuring
          creamy texture and
          preventing growth of
          ice crystals
        Suspending agent in
          soft drinks
        Puddings
        Instant desserts
        Custard tarts, etc.

E401   **Sodium alginate**

Origin   Prepared from alginic acid (E400) derived from
                  brown seaweeds.

Function   Stabilizing agent; suspending or thickening or
                  emulsifying agent in the preparation of water-
                  mixable pastes, creams and gels. Capable of
                  emulsifying an equal volume of vegetable oil by
                  simple agitation.

Adverse   No known toxicological problems.
Effects

Typical   Desserts
Products   Puddings
                  Packet cheesecake
                      mixes
                  Ice cream
                  Packet cake mixes
                  Processed cheese slices
                  Barbecue sauce mixes
                  Tinned fruit pie fillings

E402   **Potassium alginate**

Origin   Prepared from alginic acid (E400) derived from
                  native brown seaweeds.

Function   Emulsifier; stabilizer; boiler water additive; gelling
                  agent.

Adverse   None known.
Effects

Typical   —
Products

E403   **Ammonium alginate**

Origin   Prepared from alginic acid (E400) derived from
                  native brown seaweeds.

Function   Emulsifier; stab⸴ ˙ ɹer; diluent for colouring matter;
                  thickener.

Adverse   None known.
Effects

Typical   —
Products

E404   Calcium alginate (Algin)

Origin   Prepared from alginic acid (E400) derived from native brown seaweeds.

Function   Emulsifier; stabilizer; thickening agent and gelling agent.

Adverse Effects   None known.

Typical Products   Ice cream
Synthetic cream

E405   Propane -1, 2 -diol alginate (Propylene gylcol alginate; alginate ester)

Origin   Prepared from alginic acid derived from native brown seaweeds.

Function   Emulsifier or stabilizer; thickener; solvent for extracts, flavours or spices.

Adverse Effects   None known.

Typical Products   Thousand Island
    dressing
Cottage cheese with
    salmon and
    cucumber
Mint sauce
Seafood dressing
Carton salad

E406   **Agar (agar-agar; Japanese isinglass)**

Origin   A naturally occurring derivative of the stems of
                  seaweeds belonging to the red algae family,
                  especially *Gelidium amansii.*

Function   Thickening agent; stabilizer and gelling agent.

Adverse   Agar is not digested; large quantities of it may
Effects   temporarily increase flatulence and distension or
                  cause intestinal obstruction but it is likely that
                  amounts in food are too small to produce these
                  effects.

Typical   Thickening agent for
Products   ice cream and for
                      glazing meats when
                      a firm jelly is needed
                  Frozen raspberry trifle

E407   **Carrageenan (Irish Moss)**

Origin   Natural extract of several seaweeds, notably
                  Carragheen *(Chondrus crispus).*

Function   Emulsifying, thickening, suspending and gelling
                  agent.

Adverse   Reported to be the possible cause of ulcerative
Effects   colitis and, when degraded, may be carcinogenic.
                  The most harmful form is when taken in a drink
                  (Lancet, 7 Feb. '81, p. 338).

Typical   Ice creams
Products   Desserts
                  Jellified fruit juices
                  Decorations on cakes
                  Pastries
                  Biscuits
                  Blancmanges
                  Chocolate products
                  Cheeses
                  Quick-setting jelly mix
                  Milk shakes
                  Spray cream
                  Frozen trifle
                  Salad dressings
                  Sour cream
                  Infant formula
                  Alcoholic beverages

E410   **Locust bean gum (Ceratonia gum, Carob bean gum)**

Origin   Extract from the seeds of the Locust or Carob tree *Ceratonia silqua)* a member of the pea family, growing in the eastern Mediterranean.

Function   Gelling agent; stabilizer; emulsifier.

Adverse   None known. The sugary pods have been eaten
Effects   since Biblical times (and are sometimes referred to as St John's bread).

Typical   Jelly part of packet
Products   trifle mix
Italian ice cream
Tinned cherry pie
   filling
Carton salad (celery
   apple and orange)
Salad cream

E412 **Guar gum (Jaguar gum, guar flour, Cluster bean)**

Origin   A gum extracted from the stored food in the seeds of *Cyamopsis tetragonolobus*, or *C. psoraloides* a member of the pea family native to India, the drier tropics and grown in the south western areas of the United States as a cattle feed.

Function   Thickening agent; emulsion stabilizer; suspending agent; dietary bulking agent; helps diabetics control blood sugar levels.

Adverse Effects   These only occur when excessively large quantities are consumed and can include nausea, flatulence and abdominal cramps.

Typical Products
Bottled barbecue sauce
Carton salad
Scotch eggs
Salad dressings
Packet soups
Packet meringue
    crunch mix
Tinned chicken in
    white sauce
Brown sauce
Piccalilli
Horseradish cream
Sauce tartare
Carton coleslaw
Milkshake
Ice cream
Frozen fruit
Icings and glazes
Fruit drinks

E413   Tragacanth (Gum dragon, gum tragacanth)

Origin

Tragacanth gum exudes from the trunk and branches of *Astragalus gummifer* and other species of the same genera (pea family). It may flow naturally or be collected by incision in the same way as rubber is from small bushes which grow sparsely in mountain locations in Iran, Iraq, Turkey, Russia and other parts of the Middle East.

Function

Emulsifier; stabilizer; thickener; prevention of crystallization in sugar confectionery.

Adverse Effects

Adverse reactions have only occurred rarely and contact dermatitis has been reported when tragacanth was used on the skin.

Typical Products

Cottage cheese with
    salmon and
    cucumber
Cake decorations
Piccalilli
Salad dressings
Processed cheese
Cream cheese
Sherbet

E414   **Gum arabic (Acacia; Sudan gum; Gum Hashab; Kordofan gum)**

Origin   The dried gum which flows from the stems and branches of *Acacia senegal* and other African Acacia trees (members of the pea family). It collects and dries in walnut-sized globules. The trees grow in barren regions of Africa and the Middle East especially along a 3,000 km band following the southern frontier of the Sahara desert.

Function   To retard sugar crystallization; thickener; emulsifier; stabilizer and glazing agent.

Adverse Effects   A few people have demonstrated hypersensitivity to gum arabic after breathing it in or eating it.

Typical Products   Packet Black Forest
gateau mix

E415   **Xanthan gum (Corn sugar gum)**

Origin   Produced by the fermentation of a carbohydrate
                  with a bacterium called *Xanthomonas campestris.*

Function   Stabilizer; thickener; emulsifier.

Adverse   None known.
Effects

Typical   Seafood dressing
Products   Carton coleslaw and
                      carton salads
                  Horseradish cream
                  Frozen pizza
                  Packet dessert topping
                  Tinned cherry pie
                      filling
                  Sweet pickle

---

416   **Karaya Gum (Sterculia gum)**

Origin   A gum which is collected from the woody tissue
                  of members of the *Sterculiaceae* family, native
                  from southern China to Indo-China.

Function   Stabilizer; emulsifier; thickener.

Adverse   None known.
Effects

Typical   Some cheeses
Products   Fruit sauce
                  Spicy brown sauce
                  Piccalilli

131

Typical Cake Mix

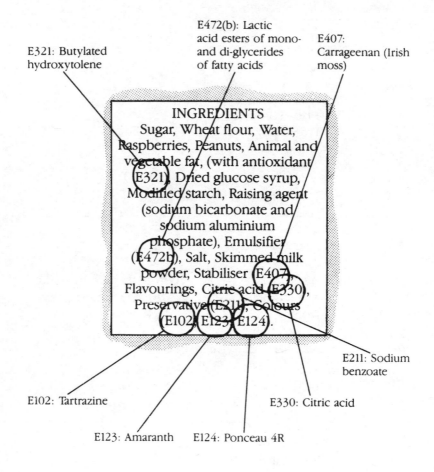

E321: Butylated hydroxytolene

E472(b): Lactic acid esters of mono- and di-glycerides of fatty acids

E407: Carrageenan (Irish moss)

INGREDIENTS
Sugar, Wheat flour, Water, Raspberries, Peanuts, Animal and vegetable fat, (with antioxidant E321), Dried glucose syrup, Modified starch, Raising agent (sodium bicarbonate and sodium aluminium phosphate), Emulsifier (E472b), Salt, Skimmed milk powder, Stabiliser (E407), Flavourings, Citric acid (E330), Preservative (E211), Colours (E102, E123, E124).

E211: Sodium benzoate

E102: Tartrazine

E330: Citric acid

E123: Amaranth   E124: Ponceau 4R

Typical Cake Mix

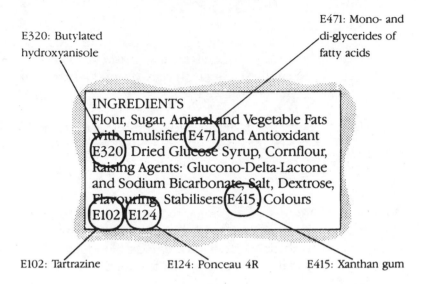

E320: Butylated
hydroxyanisole

E471: Mono- and
di-glycerides of
fatty acids

INGREDIENTS
Flour, Sugar, Animal and Vegetable Fats
with Emulsifier E471 and Antioxidant
E320 Dried Glucose Syrup, Cornflour,
Raising Agents: Glucono-Delta-Lactone
and Sodium Bicarbonate, Salt, Dextrose,
Flavouring, Stabilisers E415, Colours
E102 E124

E102: Tartrazine   E124: Ponceau 4R   E415: Xanthan gum

Typical Yogurt (Honey Muesli Flavour)

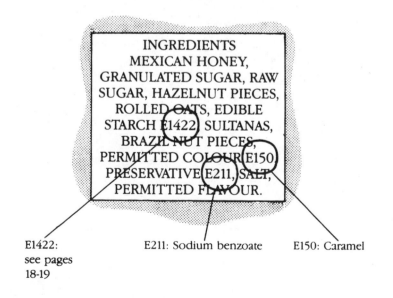

INGREDIENTS
MEXICAN HONEY,
GRANULATED SUGAR, RAW
SUGAR, HAZELNUT PIECES,
ROLLED OATS, EDIBLE
STARCH E1422 SULTANAS,
BRAZIL NUT PIECES,
PERMITTED COLOUR E150
PRESERVATIVE E211, SALT,
PERMITTED FLAVOUR.

E1422:
see pages
18-19

E211: Sodium benzoate   E150: Caramel

E420(i)   **Sorbitol**
E420(ii)   **Sorbitol syrup**

Origin   A six-carbon sugar alcohol which occurs naturally
              in some fruits and is metabolized in the body but
              is synthesized chemically from glucose for
              commercial use.

Function   Sweetening agent and substitute for glycerol.
              When added to syrups containing sucrose it
              reduces the tendency to deposit crystals on
              storage. Also as a humectant and stabilizer.

Adverse   Excessive amounts could cause flatulence,
Effects   diarrhoea and abdominal distension. Useful for
              diabetics as it does not raise the blood sugar level
              significantly, and is well tolerated.

Typical   Chocolates
Products   Confectionery
              Pastries
              Ice cream
              Food colour diluent
              Diabetic jam
              Prepackaged cakes

E421   Mannitol (Manna sugar)

Origin

Occurs naturally in the wood of coniferous trees. Generally prepared from seaweed or manna, the dried exudate of *Fraxinus ornus* which grows in Mediterranean countries.

Function

Texturizing agent; dietary supplement; humectant; sweetener in sugar-free products; anti-caking agent.

Adverse Effects

Hypersensitivity reactions have occurred occasionally and mannitol may cause nausea, vomiting and diarrhoea.

Typical Products

Sweets
Ice cream

E422   Glycerol

Origin

Occurs naturally in many plant cells, synthesized by the plants themselves. Obtained commercially from oils and fats as a by-product in the manufacture of soaps and fatty acids.

Function

Solvent (see page 18); humectant; sweetener.

Adverse Effects

Very large doses by mouth can produce effects such as headache, thirst, nausea and high blood sugar levels.

Typical Products

Liqueurs
Confectionery
Cake icing

430 Polyoxyethylene (8) stearate (Polyoxyl 8 stearate)

Origin   Manufactured from fatty acid molecules.

Function   Emulsifier; stabilizer.

Adverse Effects   Some people who have allergic skin reactions are allergic to macrogol stearate. There has been some suggestion that this additive may have an effect on the gastrointestinal and urinary tracts, forming kidney stones etc.

Typical Products   Bakery foods

Under consideration by the E.E.C. for an 'E' prefix.

431 Polyoxyethylene stearate (Polyoxyl 40 stearate)

Origin   Manufactured from fatty acids.

Function   Emulsifier.

Adverse Effects   A small proportion of people with skin allergy show allergic reactions to macrogol stearate.

Typical Products   —

Under consideration by the E.E.C. for an 'E' prefix.

432 Polyoxyethylene (20) sorbitan monolaurate (Polysorbate 20; Tween 20)

Origin Prepared from sorbitol.

Function Emulsifier; stabilizer.

Adverse Effects None known.

Typical Products —

Under consideration by the E.E.C. for an 'E' prefix.

433 Polyoxyethylene (20) sorbitan mono-oleate (Polysorbate 80; Tween 80)

Origin Prepared from sorbitol.

Function Emulsifier; stabilizer.

Adverse Effects None known.

Typical Products —

Under consideration by the E.E.C. for an 'E' prefix.

434 Polyoxyethylene (20) sorbitan monopalmitate (Polysorbate 40; Tween 40)

Origin Prepared from sorbitol.

Function Emulsifier; stabilizer.

Adverse Effects Polysorbates may increase the absorption of liquid paraffin and other fat-soluble substances.

Typical Products —

Under consideration by the E.E.C. for an 'E' prefix.

435 Polyoxyethylene (20) sorbitan monostearate (Polysorbate 60; Tween 60)

Origin Prepared from sorbitol.

Function Emulsifier; stabilizer.

Adverse Effects Polysorbates may increase the absorption of liquid paraffin and other fat-soluble substances.

Typical Products Packet cake mix

Under consideration by the E.E.C. for an 'E' prefix.

436   Polyoxyethylene (20) sorbitan tristearate (Polysorbate 65; Tween 65)

Origin   Prepared from sorbitol.

Function   Emulsifier; stabilizer.

Adverse Effects   Polysorbates may increase the absorption of liquid paraffin and other fat-soluble substances.

Typical Products   —

Under consideration by
the E.E.C. for an 'E'
prefix.

E440(a)   Pectin

Origin

Protopectin is present between the cell walls of plants, cementing them together. During the ripening process of acid fruits (especially apples, plums, bitter oranges and lemons) protopectin accumulates and as the fruit matures, enzymes break down the protopectin to the softer pectin. Apple residues from cider-making and orange pith are the commercial sources of pectin.

Function

Efficient emulsifying and gelling agent in acid media.

Adverse Effects

No real toxicological risks. Large amounts may cause temporary flatulence or intestinal distension.

Typical Products

Jams
Jellies
Marmalades
Flans
Puddings
Desserts

E440(a)   Ammonium pectate

Origin

Prepared from pectin.

Function

Stabilizer; gelling agent; thickener.

Adverse Effects

None known.

Typical Products

Jams
Jellies

E440(a)   **Potassium pectate**

Origin   Prepared from pectin.

Function   Emulsifying, stabilizing or gelling agent.

Adverse Effects   None known.

Typical Products   Jams
Preserves

E440(a)   **Sodium pectate**

Origin   Prepared from pectin.

Function   Gelling agent; emulsifier; stabilizer.

Adverse Effects   None known.

Typical Products   Jams
Preserves

E440(b)   **Amidated pectin**

Origin   Treatment of pectin extracted from citrus fruits or apples with ammonia, under alkaline conditions.

Function   Emulsifier; stabilizer; gelling agent; thickener.

Adverse Effects   None known.

Typical Products   Jams
Preserves

---

442   **Ammonium phosphatides (Emulsifier YN)**

Origin   Prepared synthetically.

Function   Stabilizer; emulsifier.

Adverse Effects   None known.

Typical Products   Cocoa and chocolate
products

Under consideration by
the E.E.C. for an 'E'
prefix.

E450(a) **_di_Sodium dihydrogen diphosphate (_di_Sodium dihydrogen pyrophosphate; acid sodium pyrophosphate)**

Origin A sodium salt of phosphoric acid.

Function Buffer; sequestrant; emulsifier; raising agent for flour; colour improver; chelating agent (see glossary p. 219).

Adverse Effects None known.

Typical Products Bread
Catering whipping
  cream
Cheese
Condensed milk and
  dried milk products

E450(a) **triSodium diphosphate**

Origin A sodium salt of phosphoric acid.

Function Buffer; sequestrant; emulsifier; raising agent for flour; colour improver; chelating agent (see glossary p. 219).

Adverse Effects None known.

Typical Products Bread
Whipping cream
  (catering)
Cheese
Condensed milk
Dried milk products

E450(a)   ***tetra*Sodium diphosphate (*tetra*Sodium pyrosphate)**

Origin   Prepared synthetically.

Function   Buffer; emulsifying salt; sequestrant; gelling agent; stabilizer.

Adverse   None known.
Effects

Typical   Cheese
Products   Catering whipping
                    cream
                 Condensed milk
                 Dried milk products
                 Frozen turkey meat loaf

E450(a)   ***tetra*Potassium diphosphate**

Origin   Prepared synthetically.

Function   Emulsifying salt.

Adverse   None known.
Effects

Typical   Instant whipped
Products   desserts
                 Packet cheesecake mix
                 Packet cake mix
                 Cheese spread
                 Ham and cooked meats

E450(b)   ***penta*Sodium triphosphate (Sodium tripolyphosphate)**

Origin   Prepared synthetically.

Function   Emulsifying salt; texturizer.

Adverse Effects   There is a French suggestion that polyphosphates could cause digestive disturbances by the blockage of a number of enzymes.

Typical Products   Cheeses
Unsweetened
condensed milk

E450(b)   ***penta*Potassium triphosphate (Potassium tripolyphosphate)**

Origin   Prepared synthetically.

Function   Emulsifying salt; texturizer.

Adverse Effects   There is a French suggestion that polyphosphates could cause digestive disturbances by the blockage of a number of enzymes.

Typical Products   Tinned hot dog
sausages

E450(c)   **Sodium polyphosphates**

Origin   Prepared synthetically.

Function   Emulsifying salts; sequestrants; stabilizers.

Adverse   There is a French suggestion that polyphosphates
Effects   could cause digestive disturbances by the
blockage of a number of enzymes.

Typical   Tinned custard
Products   Reduced sugar jam
products
Cheeses
Frozen turkey meat loaf
Frozen fish fingers and
fishcakes

E450(c)   **Potassium polyphosphates**

Origin   Prepared synthetically.

Function   Emulsifying salts; stabilizers.

Adverse   There is a French suggestion that polyphosphates
Effects   could cause digestive disturbances by the
blockage of a number of enzymes.

Typical   Cheeses
Products   Unsweetened
condensed milk
products

E460   Microcrystalline cellulose

Origin   The cellulose walls of plant fibres which are chemically fragmented into microscopic crystals.

Function   Non-nutritive bulking agent; binder; anti-caking agent; dietary fibre; hydration aid; emulsion stabilizer; heat stabilizer; alternative ingredient; tabletting binder and disintegrant; carrier and microdispersant for quick drying; cellulose component and for texture modification.

Adverse Effects   None known.

Typical Products
High-fibre bread
Low-calorie cake,
   biscuits and sweets
Reduced calorie bread
Imitation Mozzarella
   cheese
Grated and shredded
   cheese
Colours, flavours and
   food acids
Expanded snacks
Simulated fruit pieces
Imitation spices
Dehydrated foods

E460 **Alpha-cellulose (Powdered cellulose)**

Origin The cellulose component of plant cell walls which is disintegrated mechanically to form a pulp which is then dried.

Function Bulking aid; anti-caking agent; binder; dispersant; thickening agent and filter aid.

Adverse None known.
Effects

Typical —
Products
Not permitted in food
  intended specifically
  for babies or young
  children.

E461 **Methylcellulose (Methocel; cologel)**

Origin Prepared from wood pulp by treatment with alkali and methyl chloride.

Function Emulsifier; stabilizer; thickener; bulking and binding agent; film former and as a substitute for water soluble gums.

Adverse None known.
Effects

Typical Frozen bubble and
Products   squeak
Potato waffles

E463 **Hydroxypropylcellulose**

Origin Synthetically prepared ether of cellulose.

Function Stabilizer in foams and lotions; emulsifier; thickener; suspending agent.

Adverse Effects None known.

Typical Products —

E464 **Hydroxypropylmethylcellulose (Hypromellose)**

Origin Prepared from cellulose.

Function Gelling or suspending agent; emulsifier; stabilizer and thickening agent.

Adverse Effects None known.

Typical Products Frozen waffles

E465   **Ethylmethylcellulose
(Methylethylcellulose)**

Origin   Prepared from cellulose.

Function   Emulsifier; stabilizer; foaming agent.

Adverse   None known.
Effects

Typical   —
Products

E466   Carboxymethylcellulose, sodium salt (Carmellose sodium; CMC)

Origin   Prepared by treating alkali cellulose chemically.

Function   Thickening agent; texture modification; stabilizer; moisture migration control; gelling agent; non-nutritive bulking agent; opacifier etc.

Adverse Effects   One report suggests that this substance may cause intestinal obstruction, but generally safe.

Typical Products

Packet cheesecake and
   cake mixes
Icings
Bakery fillings
Fruit bar filling
Lemon pie filling
Meringues
Dips and spreads
Tinned potato salad
Tinned cream soups
Frozen whipped
   toppings
Whipped topping basis
Sterilized whipping
   cream
Ice cream
Milk shake
Frozen mousses
Tomato sauces
Salad dressings
Frozen chips
Frozen fish sticks
Batter coatings
Low-calorie orange
   squash
Processed cheese
Cottage cheese

E470 **Sodium, potassium and calcium salts of fatty acids (Soaps)**

Origin   Prepared chemically.

Function   Emulsifiers; stabilizers; anti-caking agents.

Adverse   None known.
Effects

Typical   Packet Black Forest
Products   gateau mix
                  Crispy snacks
                  Shaped crisps

E471   **Mono- and di-glycerides of fatty acids (Glyceryl monostearate, distearate)**

Origin   A normal product of digestion, but prepared for commercial use from glycerin and fatty acids.

Adverse Effects   None known.

Typical Products

Packet Black Forest
   gateau mixes
Low-cholesterol
   margarine
Quick custard mix
Hot chocolate mix
Dried potato flakes
Dehydrated mashed
   potato
Packet dessert topping
Prepacked cakes
Jam sponge pudding
   (pre-packed)
Shaped crisps
Aerosol cream
Packet savoury meal
   mix
Mousse mix

E472(a)   **Acetic acid esters of mono- and di-glycerides of fatty acids (Acetylated mono- and di-glycerides; acetoglycerides; complete and partial glycerol esters)**

Origin   Prepared from acetic acid.

Function   Emulsifiers; stabilizers; coating agents; texture modifying agents; solvents and lubricants.

Adverse Effects   None known.

Typical Products   Packet cheesecake mix
Packet dessert topping
Packet mousse mix

E472(b)   **Lactic acid esters of mono- and di-glycerides of fatty acids (Lactylated mono- and di-glycerides; Lactoglycerides)**

Origin   Prepared from lactic acid.

Function   Emulsifiers; stabilizers.

Adverse Effects   None known.

Typical Products   Packet cheesecake
    mixes
Packet dessert topping
Packet mousse mix

154

E472(c) **Citric acid esters of mono- and di-glycerides of fatty acids (Citroglycerides)**

Origin Prepared from citric acid.

Function Emulsifiers and stabilizers.

Adverse Effects None known.

Typical Products Packet dessert topping

E472(d) **Tartaric acid esters of mono- and di-glycerides of fatty acids**

Origin Prepared from tartaric acid.

Function Emulsifiers and stabilizers.

Adverse Effects None known.

Typical Products —

E472(e) **Mono- and diacetyltartaric acid esters of mono- and di-glycerides of fatty acids**

Origin Prepared from tartaric acid.

Function Emulsifiers; stabilizers.

Adverse Effects None known.

Typical Products Hot chocolate mix
Brown bread rolls
Frozen pizza
Gravy granules

E473 **Sucrose esters of fatty acids**

Origin Prepared from fatty acids.

Function Emulsifiers; stabilizers.

Adverse Effects None known.

Typical Products —

E474   **Sucroglycerides**

Origin   Prepared by the action of sucrose on natural triglycerides (from lard, tallow, palm oil etc.).

Function   Emulsifiers; stabilizers.

Adverse Effects None known.

Typical Products   —

E475   **Polyglycerol esters of fatty acids**

Origin   Prepared synthetically.

Function   Emulsifiers; stabilizers.

Adverse Effects None known.

Typical Products   Packet cheesecake and
                 cake mixes
                 Prepacked sponge
                 pudding
                 Prepacked cakes

476 **Polyglycerol esters of polycondensed fatty acids of castor oil (Polyglycerol polyricinoleate)**

Origin Prepared from castor oil.

Function Emulsifier; stabilizer.

Adverse Effects None known.

Typical Products —

Under consideration by the EEC for an 'E' prefix.

E477 **Propane -1,2 - diol esters of fatty acids (Propylene glycol esters of fatty acids)**

Origin Prepared from propylene glycol.

Function Emulsifier; stabilizer.

Adverse Effects None known.

Typical Products Packet cake mix
Instant dessert

478 Lactylated fatty acid esters of glycerol and propane -1,2-diol

Origin Prepared from lactic acid.

Function Emulsifiers, stabilizers, whipping agents, plasticizers, surface-active agents.

Adverse Effects None known.

Typical Products —

Under consideration by the E.E.C. for an 'E' prefix.

E481 Sodium stearoyl-2-lactylate

Origin Prepared from lactic acid.

Function Stabilizer; emulsifier.

Adverse Effects None known.

Typical Products
Biscuits
Bread
Cakes

Typical Soft Margarine

E322: Lecithin

E471: Mono- and di-glycerides of fatty acids

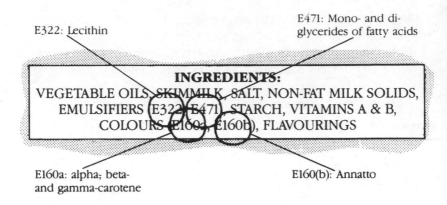

INGREDIENTS:
VEGETABLE OILS, SKIMMILK, SALT, NON-FAT MILK SOLIDS,
EMULSIFIERS (E322, E471), STARCH, VITAMINS A & B,
COLOURS (E160a, E160b), FLAVOURINGS

E160a: alpha, beta- and gamma-carotene

E160(b): Annatto

Typical Cheese Spread

E450c: Sodium polyphosphates *or* Potassium polyphosphates

E450a: *di*Sodium dihydrogen diphosphate *or tetra*sodium diphosphate *or tetra*potassium diphosphate *or tri*sodium diphosphate

E331: Sodium citrates

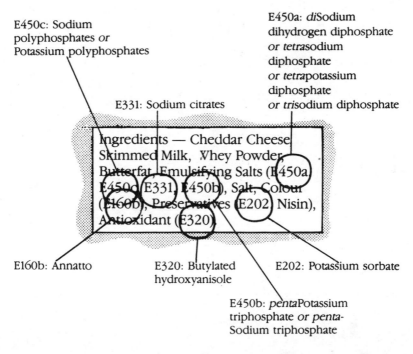

Ingredients — Cheddar Cheese, Skimmed Milk, Whey Powder, Butterfat, Emulsifying Salts (E450a, E450c, E331, E450b), Salt, Colour (E160b), Preservatives (E202, Nisin), Antioxidant (E320)

E160b: Annatto

E320: Butylated hydroxyanisole

E202: Potassium sorbate

E450b: *penta*Potassium triphosphate *or penta*-Sodium triphosphate

Typical Marmalade (Lime Flavour)

E440(a): Pectin *or*
Potassium pectate
or Sodium pectate

E142: Green S (Acid
Brilliant Green BS;
Lissamine Green)

E331: Sodium citrates

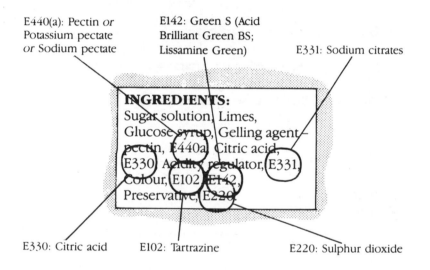

INGREDIENTS:
Sugar solution, Limes,
Glucose syrup, Gelling agent
pectin, E440a, Citric acid,
E330, Acidity regulator, E331,
Colour, E102, E142,
Preservative, E220,

E330: Citric acid

E102: Tartrazine

E220: Sulphur dioxide

E482   **Calcium stearoyl-2-lactylate**

Origin   Prepared synthetically.

Function   Emulsifier; stabilizer; whipping aid.

Adverse   None known.
Effects

Typical   Gravy granules
Products

E483   **Stearyl tartrate**

Origin   Prepared from tartaric acid.

Function   Stabilizer; emulsifier.

Adverse   None known.
Effects

Typical   —
Products

---

491   Sorbitan monostearate

Origin   Prepared synthetically from stearic acid.

Function   Emulsifier; stabilizer; glazing agent.

Adverse
Effects   None known.

Typical
Products

Packet cake mix

Under consideration by
the E.E.C. for an 'E'
prefix.

492   Sorbitan tristearate (Span 65)

Origin   Prepared synthetically from stearic acid.

Function   Emulsifier; stabilizer.

Adverse
Effects   Polysorbates may increase the body's absorption of liquid paraffin and fat-soluble substances.

Typical
Products   —

Under consideration by
the E.E.C. for an 'E'
prefix.

493   **Sorbitan monolaurate (Span 20)**

Origin   Prepared from lauric acid.

Function   Emulsifier, stabilizer, antifoaming agent.

Adverse   None known.
Effects

Typical   —
Products
Under consideration by
   the E.E.C. for an 'E'
   prefix.

---

494   **Sorbitan mono-oleate (Span 80)**

Origin   Prepared synthetically from oleic acid.

Function   Stabilizer; emulsifier.

Adverse   None known.
Effects

Typical   —
Products
Under consideration by
   the E.E.C. for an 'E'
   prefix.

495   Sorbitan monopalmitate (Span 40)

Origin   Prepared synthetically.

Function   Oil-soluble emulsifier; stabilizer.

Adverse Effects   None known.

Typical Products   —

Under consideration by
the E.E.C. for an 'E'
prefix.

500   Sodium carbonate

Origin   Although naturally occurring as saline residues deposited from the water of alkaline lakes, it is cheaper to manufacture by the Solvay process or electrolytically from sea or saline lake waters.

Function   Base.

Adverse Effects   None known in small doses. Large amounts can corrode the gut, cause gastric upsets and circulation problems.

Typical Products   Tinned custard

Under consideration by
the E.E.C. for an 'E'
prefix.

500   **Sodium hydrogen carbonate (Sodium bicarbonate; baking soda; Bicarbonate of soda)**

Origin   Prepared synthetically.

Function   Base; aerating agent; diluent.

Adverse Effects   None known.

Typical Products   Tinned custard

Under consideration by
the EEC for an 'E'
prefix.

---

500   **Sodium sesquicarbonate (Trona)**

Origin   Occurs naturally in saline residues with other minerals formed in the same way, in California, Mexico and Egypt. Prepared synthetically.

Function   Base.

Adverse Effects   None known.

Typical Products   —

Under consideration by
the E.E.C. for an 'E'
prefix.

---

501   **Potassium carbonate and Potassium hydrogen carbonate**

Origin   Prepared synthetically.

Function   Base; alkali.

Adverse Effects   None known.

Typical Products   —

Under consideration by
the E.E.C. for an 'E'
prefix.

503   **Ammonium carbonate (Hartshorn)**

Origin   Prepared synthetically.

Function   Buffer; neutralizing agent.

Adverse Effects   None known.

Typical Products   Baking powder

Under consideration by
the E.E.C. for an 'E'
prefix.

503 Ammonium hydrogen carbonate (Ammonium bicarbonate)

Origin Prepared by passing an excess of carbon dioxide through concentrated ammonia water.

Function Alkali; buffer; aerating agent.

Adverse Effects Irritates the mucous membranes of the stomach.

Typical Products —

Under consideration by the E.E.C. for an 'E' prefix.

---

504 Magnesium carbonate (Magnesite)

Origin Magnesite occurs naturally in serpentine deposits in Greece and India and replacing dolomite and limestone in Austria, Manchuria, Washington and Quebec.

Function Alkali; anti-caking agent; acidity regulator.

Adverse Effects None known.

Typical Products Table salt
Icing sugar
Soured cream
Butter
Ice cream

Under consideration by the E.E.C. for an 'E' prefix.

---

507   Hydrochloric acid

Origin   Prepared synthetically; one of the chemicals produced in the stomach to assist the digestive process.

Function   Acid.

Adverse Effects   None known.

Typical Products   —

Under consideration by the E.E.C. for an 'E' prefix.

508   Potassium chloride

Origin   Occurs naturally as a saline residue associated with rock salt, and around volcanic vents.

Function   Gelling agent; salt substitute; dietary supplement.

Adverse Effects   Can, in large doses, cause intestinal ulceration, sometimes with haemorrhage and perforation. Gastric ulceration may occur with sustained release tablets. Unpleasant taste in solution can cause nausea and vomiting.

Typical Products   Salt substitute

Under consideration by the E.E.C. for an 'E' prefix.

509   **Calcium chloride**

Origin

Obtained as a by-product of the solvay process and is also a product from natural salt brines.

Function

Sequestrant; firming agent.

Adverse Effects

None known.

Typical Products

Tinned red kidney
     beans
Pickled red cabbage

Under consideration by
     the E.E.C. for an 'E'
     prefix.

---

510   **Ammonium chloride**

Origin

Prepared synthetically.

Function

Yeast food; flavour.

Adverse Effects

Ammonium chloride is readily absorbed by the food canal and may decrease the acidity of the urine. It should be avoided by people with imperfect liver or kidney functions.

Typical Products

—

Under consideration by
     the E.E.C. for an 'E'
     prefix.

---

513   Sulphuric acid

Origin   Prepared synthetically.

Function   Acid.

Adverse
Effects   Poisonous.

Typical
Products   —

Under consideration by
the E.E.C. for an 'E'
prefix.

514   Sodium sulphate

Origin   Occurs naturally as thenardite and mirabilite, and the U.S.S.R. Canada and the U.S.A. are the chief producers.

Function   Diluent.

Adverse
Effects   Although the healthy body can adapt to a wide range of sodium intake daily, excessive sodium can be dangerous because it is closely related to the body's water balance. Those at greatest risk are small babies and people suffering with kidney and heart complaints.

Typical
Products   —

Under consideration by
the E.E.C. for an 'E'
prefix.

515   Potassium sulphate

Origin

Occurs in nature as a triple sulphate of potassium magnesium and calcium particularly at Stassfurt in Germany.

Function

Salt substitute for dietetic use.

Adverse Effects

None known.

Typical Products

—

Under consideration by the E.E.C. for an 'E' prefix.

516   Calcium sulphate (Gypsum, Plaster of Paris)

Origin

A naturally occurring mineral; commercial sources are the U.S.A. and France, followed by Spain, Great Britain and Canada.

Function

Firming agent; sequestrant; nutrient; yeast food; inert excipient.

Adverse Effects

None known.

Typical Products

—

Under consideration by the E.E.C. for an 'E' prefix.

518   **Magnesium sulphate — (Epsom salts, Espomite)**

Origin   Occurs in solution in sea- and mineral-waters and is deposited from the waters of saline lakes and as crusts in limestone caves.

Function   Dietary supplement; firming agent; used in making 'Burton' style beer.

Adverse Effects   Magnesium is not absorbed to any large extent by the body so that toxicity is not a problem except to people whose kidneys are functioning imperfectly.

Typical Products   —

Under consideration by the E.E.C. for an 'E' prefix.

524   **Sodium hydroxide**

Origin   Prepared synthetically.

Function   Base; colour solvent.

Adverse Effects   None known.

Typical Products   Jams and preserves

Under consideration by the E.E.C. for an 'E' prefix.

525   Potassium hydroxide

Origin   Manufactured synthetically.

Function   Base.

Adverse Effects   Unless it is very dilute, the caustic nature of this chemical causes immediate burning pain in the mouth, throat and stomach and the lining membranes become swollen and detached. Vomiting, pain and shock result.

Typical Products   Cocoa products

Under consideration by the E.E.C. for an 'E' prefix.

526   Calcium hydroxide

Origin   Prepared by the hydration of lime.

Function   Firming agent, neutralizing agent.

Adverse Effects   None known.

Typical Products   Cheese
Cocoa products
Shaped crisps

Under consideration by the E.E.C. for an 'E' prefix.

527   **Ammonium hydroxide**

Origin   Prepared synthetically.

Function   Food colouring diluent and solvent; alkali.

Adverse   None known.
Effects

Typical   Food colours
Products   Cocoa products

                 Under consideration by
                    the E.E.C. for an 'E'
                    prefix.

---

528   **Magnesium hydroxide**

Origin   Occurs in nature as the mineral periclase. It is
                 prepared commercially from magnesite ores.

Function   Alkali.

Adverse   None known.
Effects

Typical   Cocoa products
Products
                 Under consideration by
                    the E.E.C. for an 'E'
                    prefix.

---

529   Calcium oxide

Origin   Prepared from limestone.

Function   Alkali; nutrient.

Adverse   None known.
Effects

Typical   Some cocoa products.
Products

Under consideration by
    the E.E.C. for an 'E'
    prefix.

530   Magnesium oxide
(Periclase; Native magnesium)

Origin   A naturally occurring mineral particularly in rocks which have undergone change brought about by pressure and heat.

Function   Anti-caking agent; alkali.

Adverse   None known.
Effects

Typical   Some cocoa products
Products

Under consideration by
    the E.E.C. for an 'E'
    prefix.

535   **Sodium ferrocyanide**
             (Sodium hexacyanoferrate II)

Origin   Manufactured synthetically.

Function   Anti-caking agent; crystal modifier.

Adverse   There is a very strong chemical bondage between
Effects   the iron and cyanide groups which prevents
             ferrocyanides from having a high order of toxicity.

Typical   —
Products
             Under consideration by
                the E.E.C. for an 'E'
                prefix.

536   **Potassium ferrocyanide**
             (Potassium hexacyanoferrate II)

Origin   Prepared synthetically.

Function   Anti-caking agent.

Adverse   Because the iron and cyanide groups are strongly
Effects   bonded there is a very low level of toxicity.

Typical   Some wines
Products
             Under consideration by
                the E.E.C. for an 'E'
                prefix.

177

540   *di*Calcium diphosphate (Calcium hydrogen phosphate; Calcium phosphate dibasic)

Origin   Occurs in nature as the mineral monetite, also prepared synthetically.

Function   Mineral supplement in cereals and other foods. Buffer; neutralizing agent; dietary supplement; raising agent.

Adverse Effects   Little *di*calcium diphosphate is absorbed by the intestines and there is little danger of any adverse reaction.

Typical Products   Some cheeses
Shaped crisps

Under consideration by the E.E.C. for an 'E' prefix.

541   Sodium aluminium phosphate, acidic

Origin   Prepared synthetically.

Function   Acid; raising agent for flour.

Adverse Effects   Although the healthy body can adapt to a wide range of sodium intake daily, excessive sodium can be dangerous because it is closely related to the body's water balance. Those at greatest risk are small babies and people suffering with kidney and heart complaints.

*Typical
Products*

Packet cake mixes

Under consideration by
the E.E.C. for an 'E'
prefix.

541 Sodium aluminium phosphate, basic

Origin   Prepared synthetically.

Function   Emulsifying salt.

*Adverse
Effects*   None known.

*Typical
Products*   —

Under consideration by
the E.E.C. for an 'E'
prefix.

542 Edible bone phosphate

Origin   The degreased steam-extract from animal bones.

Function   Anti-caking agent; mineral supplement; filler in
tablet making.

*Adverse
Effects*   None known.

*Typical
Products*   —

Under consideration by
the E.E.C. for an 'E'
prefix.

544   Calcium polyphosphates

Origin

Prepared synthetically.

Function

Emulsifying salts — have an action on milk proteins which prevents processed cheese from separating out.

Adverse Effects

There is some suggestion that polyphosphates could cause digestive disturbances by the blocking a number of enzymes.

Typical Products

Cheeses

Under consideration by the E.E.C. for an 'E' prefix.

545   Ammonium polyphosphates

Origin

Prepared synthetically.

Function

Emulsifiers; emulsifying salts; curing aids; water-binding aids. Enables manufacturer to add water to frozen chicken.

Adverse Effects

There is some suggestion that polyphosphates could cause digestive disturbances by the blocking a number of enzymes.

Typical Products

Cheeses

Under consideration by the E.E.C. for an 'E' prefix.

551   **Silicon dioxide (Silicea, Silica)**

Origin   Silicon dioxide is the commonest rock-forming
                   mineral and sand is composed mainly of small
                   grains of quartz or flint, both of which are silicon
                   dioxide. In the food industry the grains are further
                   processed into a microcellular powder producing
                   a gel form and a colloidal form by further
                   hydrolysis.

Function   Suspending and anti-caking agent; thickener and
                   stabilizer in suspensions and emulsions, including
                   wine.

Adverse   None known.
Effects

Typical   Shaped crisps
Products
                   Under consideration by
                      the E.E.C. for an 'E'
                   prefix.

552   Calcium silicate

Origin   A naturally occurring mineral. Commercial calcium silicate is prepared synthetically.

Function   Anti-caking agent; in pharmacology as an antacid; glazing, polishing and release agent (sweets); dusting agent (chewing gum); coating agent (rice).

Adverse   None known.
Effects

Typical   Salt
Products   Garlic and onion salt
              Icing sugar
              Sweets
              Rice
              Chewing gum

              Under consideration by
                 the E.E.C. for an 'E'
                 prefix.

553(a)   **Magnesium silicate, synthetic and Magnesium trisilicate**

Origin

Magnesium silicate is a synthetic compound of magnesium oxide, and silicon dioxide. Magnesium trisilicate occurs in nature as the minerals meerschaum, parasepiolite and sepiolite.

Function

Anti-caking agent and tablet excipient and as an antacid in pharmacology; glazing, polishing and release agent (sweets); dusting agent (chewing gum); coating agent (rice).

Adverse Effects

None known.

Typical Products

Salt
Garlic and onion salt
Icing sugar
Sweets
Rice
Chewing gum

Under construction by
the E.E.C. for an 'E'
prefix.

553(b)　Talc (French chalk)

Origin　A naturally occurring mineral, worked in the U.S.A., France, Italy, Canada etc.

Function　Release agent.

Adverse Effects　None known.

Typical Products　—

Under consideration by
　the E.E.C. for an 'E'
　prefix.

554　Aluminium sodium silicate

Origin　Naturally occurring mineral, known as analcite and natrolite.

Function　Anti-caking agent.

Adverse Effects　None known.

Typical Products　Packet noodles

Under consideration by
　the E.E.C. for an 'E'
　prefix.

556   **Aluminium calcium silicate**
        (Calcium aluminium silicate)

Origin   Naturally occurring mineral, known as scolecite
and heulandite.

Function   Anti-caking agent.

Adverse   None known.
Effects

Typical   —
Products

     Under consideration by
       the E.E.C. for an 'E'
       prefix.

558   **Bentonite (Bentonitum; Soap clay)**

Origin   A particular clay deposit occurring in thin beds
in the western U.S.A., believed to result from the
decomposition of volcanic ash.

Function   Anti-caking agent; clarifying agent; suspending
and emulsifying agent.

Adverse   None known.
Effects

Typical   —
Products

     Under consideration by
       the E.E.C. for an 'E'
       prefix.

559   Kaolin, heavy and Kaolin, light

Origin

Occurs in nature as an altered mineral in granite, particularly in Cornwall, the U.S.A., France, China and Malaya.

Function

Anti-caking agent.

Adverse Effects

None known.

Typical Products

—

Under consideration by the E.E.C. for an 'E' prefix.

570   Stearic acid

Origin

Naturally occurring fatty acid found in all animal fats and vegetable oils. Prepared synthetically for commercial use.

Function

Anti-caking agent.

Adverse Effects

None known.

Typical Products

—

Under consideration by the E.E.C. for an 'E' prefix.

572   **Magnesium stearate**

Origin   Prepared synthetically from commercial stearic acid.

Function   Anti-caking agent; emulsifier; release agent.

Adverse Effects   None are known from the consumption of this additive but accidental inhalation of the powder can be harmful.

Typical Products   Sweets made by direct compression

Under consideration by the E.E.C. for an 'E' prefix.

575   **D-Glucono-1,5-lactone, (Glucono *delta*-lactone)**

Origin   Prepared by oxidation of glucose.

Function   Acid; sequestrant. In dairy industry prevents milkstone formation (deposits of magnesium and calcium phosphates etc., when milk is heated to a high temperature); in breweries prevents beerstone formation.

Adverse Effects   None known.

Typical Products   Packet cake mix

Under consideration by the E.E.C. for an 'E' prefix.

Typical Fish Fingers

E124: Ponceau 4R   E102: Tartrazine

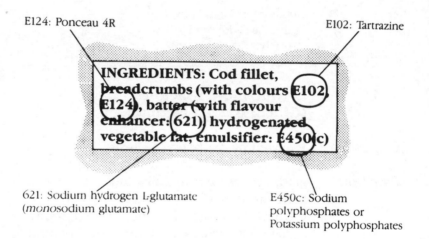

INGREDIENTS: Cod fillet, breadcrumbs (with colours E102, E124), batter (with flavour enhancer: 621) hydrogenated vegetable fat, emulsifier: E450(c)

621: Sodium hydrogen L-glutamate (monosodium glutamate)

E450c: Sodium polyphosphates or Potassium polyphosphates

188

576   **Sodium gluconate**

Origin   Prepared synthetically; the sodium salt of gluconic acid.

Function   Sequestrant; dietary supplement.

Adverse   None known.
Effects

Typical   —
Products
                  Under consideration by
                  the E.E.C. for an 'E'
                  prefix.

---

577   **Potassium gluconate**

Origin   Prepared synthetically; the potassium salt of gluconic acid.

Function   Sequestrant.

Adverse   None known.
Effects

Typical   —
Products
                  Under consideration by
                  the E.E.C. for an 'E'
                  prefix.

578   Calcium gluconate

Origin   Prepared synthetically; calcium salt prepared from gluconic acid.

Function   Buffer; firming agent; sequestrant.

Adverse Effects   None known.

Typical Products   —

Under consideration by
the E.E.C. for an 'E'
prefix.

620   L-glutamic acid

Origin   A naturally occurring amino-acid of great importance in the nitrogen metabolism of plants and animals, but prepared commercially by the fermentation of a carbohydrate solution by a bacterium e.g. *Micrococcus glutamicus*. Several other methods exist.

Function   Dietary supplement; flavour enhancer; salt substitute.

Adverse Effects   None known.

Typical Products   —

Under consideration by
the E.E.C. for an 'E'
prefix.

621   **Sodium hydrogen L-glutamate (*mono*Sodium glutamate; Aji-no-moto; MSG)**

Origin   Occurs naturally in a Japanese seaweed called Seatango; prepared commercially from sugar beet pulp and wheat gluten.

Function   Flavour-enhancer of proteinaceous foods either by increasing the amount of saliva produced in the mouth or by stimulating the taste buds.

Adverse   Can give rise to a condition known as 'Chinese
Effects   Restaurant syndrome' the symptoms of which are heart palpitations, headaches, dizziness, muscle tightening, nausea, weakness of the upper arms, pains in the neck and symptoms similar to migraine in some people. Prohibited in or on foods intended for babies or young children. (See pages 12-14.)

Typical   Packet snacks
Products   Chilli sauce
                 Frozen potato waffles
                 Pork pies
                 Pork sausages
                 Packet soup and quick
                     soups
                 Flavoured noodles

                 Under consideration by
                     the E.E.C. for an 'E'
                     prefix.

---

622   **Potassium hydrogen L-glutamate (*mono*Potassium glutamate)**

Origin   Prepared synthetically.

Function   Flavour enhancer; salt substitute.

Adverse Effects   Sometimes nausea, vomiting, diarrhoea and abdominal cramps may occur, although there is usually little toxicity of potassium salts when taken by mouth in healthy individuals as potassium is rapidly excreted in the urine. Potassium could be harmful for those with impaired kidneys. Not to be given to babies under 12 weeks old. (See also pages 12-14.)

Typical Products   Sodium-free
        condiments

        Under consideration by
        the E.E.C. for an 'E'
        prefix.

---

623   **Calcium dihydrogen di-L-glutamate (Calcium glutamate)**

Origin   Prepared synthetically.

Function   Flavour enhancer; salt substitute.

Adverse Effects   None known, but not to be given to babies under 12 weeks old. (See also pages 12-14.)

Typical Products   Dietetic foods

        Under consideration by
        the E.E.C. for an 'E'
        prefix.

---

627 **Guanosine 5'-(disodium phosphate) (Sodium guanylate)**

Origin   The sodium salt of 5' Guanylic acid, a widely-occurring nucleotide (isolated from sardines and yeast extract), prepared synthetically for commercial use.

Function   Flavour-enhancer.

Adverse   None known, but prohibited in or on foods
Effects   intended for babies or young children. People suffering from conditions such as gout, which require the avoidance of purines are recommended to avoid this substance. (See also pages 12-14.)

Typical   Pre-cooked dried rice
Products   snacks
               Crisps
               Gravy granules

               Under consideration by
                  the E.E.C. for an 'E'
                  prefix.

631   **Inosine 5'-(disodium phosphate) (Sodium 5'-inosinate)**

Origin   The disodium salt of inosinic acid, (muscle inosinic acid) which can be prepared from meat extract and dried sardines.

Function   Flavour enhancer.

Adverse Effects   None known, but prohibited in or on foods specially made for babies and young children. People suffering from conditions such as gout, which require the avoidance of purines, should avoid this substance. (See also pages 12-14.)

Typical Products   Pre-cooked dried rice
        snacks
Some crisps
Gravy granules

Under consideration by
     the E.E.C. for an 'E'
     prefix.

---

635   **Sodium 5'-ribonucleotide**

Origin   A mixture of disodium guanylate and disodium inosinate.

Function   Flavour enhancer.

Adverse Effects   None known but not permitted in foods specially prepared for babies or young children. People suffering from conditions such as gout, which require the avoidance of purines, should avoid this substance. (See also pages 12-14.)

---

*Typical
Products*

Frozen croquette
    potatoes
Potato waffles
Mini waffles

Under consideration by
    the E.E.C. for an 'E'
    prefix.

636   Maltol

Origin

A naturally occurring substance found in the bark of young larch trees, pine needles, chicory, wood tars, oils and roasted malt. Also obtained chemically by alkaline hydrolysis of streptomycin salts.

Function

Flavouring agent, to impart 'freshly baked' smell and flavour to bread and cakes.

*Adverse
Effects*

None known.

*Typical
Products*

Bread
Cakes

Under consideration by
    the E.E.C. for an 'E'
    prefix.

637   **Ethyl maltol**

Origin   Chemically prepared from maltol.

Function   Flavouring, to impart a sweet taste.

Adverse   None known.
Effects

Typical   —
Products

               Under consideration by
                 the E.E.C. for an 'E'
                 prefix.

---

900   Dimethylpolysiloxane (Simethicone; Dimethicone)

Origin

A chemically manufactured mixture of liquid dimethylpolysiloxane and silica gel or silicon dioxide.

Function

Water repellent; anti-foaming agent.

Adverse Effects

None known.

Typical Products

Jams
Pineapple juice

Products containing dimethylpolysiloxane may also contain formaldehyde in proportion to the additives use up to 1000mg/kg.

Under consideration by the E.E.C. for an 'E' prefix.

---

901   Beeswax, white and Beeswax, yellow

Origin   A naturally occurring product from the bee honeycomb. White beeswax is the bleached and purified form.

Function   Glazing and polishing agent, release agent.

Adverse Effects   None known.

Typical Products   Diluents for food
     colours

Under consideration by
the E.E.C. for an 'E'
prefix.

---

903 Carnauba wax

Origin

A yellow to light brown wax obtained from the surface of leaves of *Copernicia cerifera*, the Brazilian wax palm.

Function

Glazing and polishing agent for sugar confectionery.

Adverse Effects

None known.

Typical Products

Only permitted in
    chocolate products
    such as in:
Sugar confectionery
Chocolate
    confectionery

Under consideration by
    the E.E.C. for an 'E'
    prefix.

---

904   **Shellac**

Origin   Shellac is a substance obtained from the resin
produced by the Lac insect (*Laccifer lacca,* a
member of the *Lacciferidae*) related to mealy bugs
and scale insects belonging to the *Coccoidea*. It
is a native of India. Four commercial grades are
produced by different chemical processes.

Function   Glazing agent.

Adverse   No significant reports of adverse effects.
Effects

Typical   Cake decorations
Products   Sweets
Fizzy orange drink
Sugar strands

Under consideration by
the E.E.C. for an 'E'
prefix.

Mineral Hydrocarbons

905   **Mineral hydrocarbons**

Origin   Chemically manufactured mineral derivatives.

Function   Polishes, glazing agents, sealing agents, ingredient for chewing gum.

Adverse   Excessive dosage may result in anal seepage and
Effects   irritation.

Typical   Dried fruit (to prevent
Products   sugaring of berries
                        and clouding of film
                        bags)
                    Citrus fruit
                    Sugar confectionery
                    Chewing gum
                    Cheese rind
                    Eggs (through dipping
                        or spraying for
                        preserving which
                        must be declared)

                    Under consideration by
                        the E.E.C. for an 'E'
                        prefix.

907   Refined microcrystalline wax

Origin   Prepared by solution of the heavy fraction of petroleum by dewaxing or deoiling methods.

Function   Chewing gum ingredient; polishing and release agent. Also stiffening agent and for tablet coating.

Adverse Effects   None known.

Typical Products   —

Under consideration by
  the E.E.C. for an 'E'
  prefix.

---

920   L-cysteine hydrochloride and L-cysteine hydrochloride monohydrate

Origin   A synthetically prepared derivative of the naturally occurring amino-acid, cysteine.

Function   Improving agent for flour, other than wholemeal.

Adverse Effects   None known.

Typical Products   Flour and bakery
  products

Under consideration by
  the E.E.C. for an 'E'
  prefix.

924	**Potassium bromate**

Origin   Prepared synthetically.

Function   Oxidizing agent; flour maturing agent; flour bleaching agent.

Adverse Effects   Can cause nausea, vomiting, severe abdominal pain, diarrhoea and even convulsions. Bleaching of flour destroys much of its natural vitamin E and other nutrients and the chemicals used have not all been proved to be completely safe.

Typical Products   Bread

Under consideration by
the E.E.C. for an 'E'
prefix.

925	**Chlorine**

Origin   Prepared synthetically.

Function   Antibacterial and antifungal preservative; bleaching agent.

Adverse Effects   Powerful irritant. Bleaching of flour has never been demonstrated to be 100 per cent safe. The process takes its toll of flour nutrients and destroys much of the vitamin E content.

Typical Products   Flour

Under consideration by
the E.E.C. for an 'E'
prefix.

925   Chlorine dioxide

Origin   Prepared synthetically.

Function   Bleaching and improving agent for flour; bleaching agent for fats and oils, beeswax etc; purification of water; taste and odour control of water; oxidizing agent; bactericide and antiseptic.

Adverse Effects   Bleaching of flour has never been demonstrated to be 100 per cent safe. It takes its toll of flour nutrients and destroys much of the vitamin E.

Typical Products   Flour

Under consideration by
the E.E.C. for an 'E'
prefix.

927   Azodicarbonamide (Azoformamide)

Origin   Prepared synthetically.

Function   Flour improving agent to improve the tolerance of bread dough under a wide range of fermentation conditions.

Adverse Effects   None known.

Typical Products   Flour

Under consideration by
the E.E.C. for an 'E'
prefix.

Alphabetical List of Additives and Their 'E' Numbers

E414	Acacia
E260	Acetic acid
E472(a)	Acetic acid esters of mono- and di-glycerides of fatty acids
E472(e)	mono- and diacetyltartaric esters of mono- and di-glycerides of fatty acids
E142	Acid Brilliant Green BS
355	Adipic acid
E406	Agar
621	Aji-no-moto
E405	Alginate ester
E400	Alginic acid
E160(a)	Alpha-carotene
E460	*Alpha*-cellulose
E173	Aluminium
556	Aluminium calcium silicate
554	Aluminium sodium silicate
E123	Amaranth
E440(b)	Amidated pectin
E403	Ammonium alginate
503	Ammonium carbonate
510	Ammonium chloride
380	*tri*-Ammonium citrate
381	Ammonium ferric citrate
381	Ammonium ferric citrate; green
503	Ammonium hydrogen carbonate
527	Ammonium hydroxide
442	Ammonium phosphatides

545	Ammonium polyphosphates
E160(b)	Annatto
E163	Anthocyanins
E300	L-Ascorbic acid
E304	Ascorbyl palmitate
927	Azodicarbonamide
927	Azoformamide
E122	Azorubine
E162	Beetroot Red
901	Beeswax, white
901	Beeswax, yellow
E160(e)	beta-apo-8'-carotenal
E160(a)	alpha-carotene
558	Bentonite
E210	Benzoic acid
E162	Betanin
E320	BHA
E321	BHT
500	Bicarbonate of soda
E230	Biphenyl
E161(b)	Bixin
E151	Black PN
133	Brilliant blue FCF
154	Brown FK
E320	Butylated hydroxyanisole
E321	Butylated hydroxytoluene
E263	Calcium acetate
E404	Calcium alginate
E302	Calcium L-ascorbate
E213	Calcium benzoate
E227	Calcium bisulphite
E170	Calcium carbonate
509	Calcium chloride
509	Calcium chloride; anhydrous
E333	monoCalcium citrate
E333	diCalcium citrate
E333	triCalcium citrate
E238	Calcium formate

578	Calcium gluconate
623	Calcium glutamate
623	Calcium di-hydrogen di-L-glutamate
352	Calcium hydrogen malate
E341(b)	Calcium hydrogen orthophosphate
E540	Calcium hydrogen phosphate
E341(b)	Calcium tetrahydrogen diorthophosphate
529	Calcium oxide
E227	Calcium hydrogen sulphite
526	Calcium hydroxide
E327	Calcium lactate
352	Calcium malate
E341(c)	*tri*Calcium diorthophosphate
540	*di*Calcium diphosphate
544	Calcium polyphosphates
E282	Calcium propionate
552	Calcium silicate
385	Calcium disodium EDTA
385	Calcium disodium ethylenediamine-NNN'N' tetra-acetate
E203	Calcium sorbate
E482	Calcium stearoyl-2-lactylate
516	Calcium sulphate
E226	Calcium sulphite
E161(g)	Canthaxanthin
E160(c)	Capsanthin
E160(c)	Capsorubin
E150	Caramel
E153	Carbon black
E290	Carbon dioxide
E466	Carboxymethylcellulose, sodium salt
E120	Carmine of Cochineal
E120	Carminic acid
E122	Carmoisine
903	Carnauba wax
E410	Carob gum
E160(a)	Carotene, alpha, beta and gamma
E407	Carrageenan

E465	Cellulose, ethyl methyl
E463	Cellulose, hydroxypropyl
E464	Cellulose, hydroxypropylmethyl
E460(i)	Cellulose, microcrystalline
E460(ii)	Cellulose, powdered
925	Chlorine
926	Chlorine dioxide
E140	Chlorophyll
154	Chocolate brown FK
155	Chocolate brown HT
E330	Citric acid
E472(c)	Citric acid esters of mono- and di-glycerides of fatty acids
E472(c)	Citroglycerides
E466	CMC
E120	Cochineal
E124	Cochineal Red A
E461	Cologel
E336	Cream of tartar
E161(c)	Cryptoxanthin
E100	Curcumin
920	L-cysteine hydrochloride
920	L-cysteine hydrochloride monohydrate
E472(e)	Diacetyltartaric acid esters of fatty acids
E333	diCalcium citrate
540	diCalcium diphosphate
900	Dimethylpolysiloxane
E230	Diphenyl
E340(b)	diPotassium hydrogen orthophosphate
E336	diPotassium L-(+)-tartrate
E450(a)	diSodium dihydrogen diphosphate
E339(a)	diSodium hydrogen orthophosphate
E335	diSodium L-(+)-tartrate
E312	Dodecyl gallate
542	Edible bone phosphate
442	Emulsifier YN
E127	Erythrosine BS
E160(f)	Ethyl ester of beta-apo-8'-carotenoic acid

E214	Ethyl-4-hydroxybenzoate
E214	Ethyl *para*-hydroxybenzoate
E215	Ethyl-4-hydroxybenzoate, sodium salt
637	Ethyl maltol
E465	Ethylmethylcellulose
E306	Extracts of natural origin rich in tocopherols
E161(a)	Flavoxanthin
E236	Formic acid
553(b)	French chalk
297	Fumaric acid
E414	Gum Arabic
E160(a)	gamma-carotene
575	Glucono *delta*-lactone
575	D-Glucono-1, 5-lactone
620	L-glutamic acid
E471	Mono- and di-glycerides of fatty acids
E422	Glycerol
E471	Glyceryl monostearate
E197	Gold
E142	Green S
627	Guanosine 5^2 (disodium phosphate)
E412	Guar gum
E414	Gum arabic
370	1,4-Heptonolactone
E239	Hexamine
E239	Hexamethylenetetramine
507	Hydrochloric acid
E231	2-Hydroxybiphenyl
E463	Hydroxypropylcellulose
E464	Hydroxypropylmethylcellulose
E464	Hypromellose
E132	Indigo carmine
631	Inosine 5^2 (disodium phosphate)
E172	Iron hydroxides
E172	Iron oxides
559	Kaolin
416	Karaya gum
154	Kipper Brown
E270	Lactic acid

E472(b)	Lactic acid esters of mono- and di-glycerides of fatty acids
E101	Lactoflavin
E472(b)	Lactoglycerides
478	Lactylated fatty acid esters of glycerol and propane -1, 2-diol
E322	Lecithin
620	L-glutamic acid
E142	Lissamine Green
E180	Lithol Rubine BK
E410	Locust bean gum
E161(b)	Lutein
E160(d)	Lycopene
504	Magnesium carbonate
528	Magnesium hydroxide
530	Magnesium oxide
553(a)	Magnesium silicate synthetic
553(a)	Magnesium trisilicate
572	Magnesium stearate
518	Magnesium sulphate
296	DL-Malic acid, L-Malic acid
636	Maltol
E163	Malvidin
E421	Mannitol
353	Metatartaric acid
E461	Methocel
E461	Methylcellulose
E218	Methyl 4-hydroxybenzoate
E218	Methyl *para*-hydroxybenzoate
E219	Methyl 4-hydroxybenzoate, sodium salt
E460(i)	Microcrystalline cellulose
907	Microcrystalline wax, refined
905	Mineral hydrocarbons
E472(e)	Mono- and diacetyltartaric acid esters of mono- and di-glycerides of fatty acids
E471	Mono- and di-glycerides of fatty acids
E333	*mono*Calcium citrate
E332	*mono*Potassium citrate

622	*mono*Potassium glutamate
E336	*mono*Potassium L-(+)-tartrate
E331	*mono*Sodium citrate
621	*mono*Sodium glutamate
E335	*mono*Sodium L-(+)-tartrate
621	MSG
375	Niacin
375	Nicotinic acid
234	Nisin
E160(b)	Norbixin
570	Octadecarnoic acid
E311	Octyl gallate
E338	Orthophosphoric acid
E231	Orthophenylphenol
E304	6-0-Palmitoyl-L-ascorbic acid
E131	Patent blue V
E440(a)	Pectin
E163	Pelargonidin
E450(b)	*penta*Potassium triphosphate
E450(b)	*penta*Sodium triphosphate
E163	Peonidin
E163	Petunidin
E338	Phosphoric acid
E180	Pigment Rubine
E475	Polyglycerol esters of fatty acids
476	Polyglycerol esters of polycondensed fatty acids of castor oil
476	Polyglycerol polyricinoleate
432	Polyoxyethylene (20) sorbitan monolaurate
433	Polyoxyethylene (20) sorbitan mono-oleate
434	Polyoxyethylene (20) sorbitan monopalmitate
435	Polyoxyethylene (20) sorbitan monostearate
436	Polyoxyethylene (20) sorbitan tristearate
430	Polyoxyethylene (8) stearate
431	Polyoxyethylene (40) stearate
544	Polyphosphates, calcium
545	Polyphosphates, ammonium
E450(c)	Polyphosphates, potassium and sodium
432	Polysorbate 20

434	Polysorbate 40
435	Polysorbate 60
436	Polysorbate 65
433	Polysorbate 80
E124	Ponceau 4R
E261	Potassium acetate
E402	Potassium alginate
E212	Potassium benzoate
924	Potassium bromate
501	Potassium carbonate
508	Potassium chloride
E332	*mono*Potassium citrate
E332	*tri*Potassium citrate
E450(a)	*tetra*Potassium diphosphate
536	Potassium ferrocyanide
577	Potassium gluconate
622	*mono*Potassium glutamate
536	Potassium hexacyanoferrate (II)
501	Potassium hydrogen carbonate
E332	Potassium dihydrogen citrate
622	Potassium hydrogen L-glutamate
E340(a)	Potassium dihydrogen orthophosphate
E340(b)	*di*Potassium hydrogen orthophosphate
E336	Potassium hydrogen tartrate
525	Potassium hydroxide
E326	Potassium lactate
351	Potassium malate
E224	Potassium metabisulphite
E252	Potassium nitrate
E249	Potassium nitrite
E340(c)	*tri*Potassium orthophosphate
E440(a)	Potassium pectate
450(a)	*tetra*Potassium diphosphate
450(b)	*penta*Potassium triphosphate
E450(c)	Potassium polyphosphates
E283	Potassium propionate
E470	Potassium salts of fatty acids

E337	Potassium sodium DL-tartrate
E337	Potassium sodium L-(+)-tartrate
E202	Potassium sorbate
515	Potassium sulphate
E336	*mono*Potassium L-(+)-tartrate
E336	*di*Potassium L-(+)-tartrate
E450(b)	penta Potassium triphosphate
E460(ii)	Powdered cellulose
E405	Propane -1, 2-diol alginate
E477	Propane -1, 2-diol esters of fatty acids
E280	Propionic acid
E405	Propylene glycol alginate
E310	Propyl gallate
E216	Propyl 4-hydroxybenzoate
E216	Propyl *para*-hydroxybenzoate
E217	Propyl 4-hydroxybenzoate, sodium salt
E104	Quinoline yellow
128	Red 2G
907	Refined microcrystalline wax
E161(f)	Rhodoxanthin
E101	Riboflavin
101(a)	Riboflavin-5'-phosphate
E180	Rubine
E161(d)	Rubixanthin
904	Shellac
551	Silica
551	Silicon dioxide
E174	Silver
E262	Sodium acetate
E401	Sodium alginate
541	Sodium aluminium phosphate
E301	Sodium L-ascorbate
E211	Sodium benzoate
500	Sodium bicarbonate
E232	Sodium biphenyl-2-yl oxide
E222	Sodium bisulphite
500	Sodium carbonate
E466	Sodium carboxymethylcellulose
E331	*mono*Sodium citrate

E331	*di*Sodium citrate
E331	*tri*Sodium citrate
E450(a)	*tetra* Sodium diphosphate
E450(a)	*tri*Sodium diphosphate
E215	Sodium ethyl *para*-hydroxybenzoate
535	Sodium ferrocyanide
E237	Sodium formate
576	Sodium gluconate
621	*mono*Sodium glutamate
627	Sodium guanylate
535	Sodium hexacyanoferrate (II)
500	Sodium hydrogen carbonate
E331	Sodium dihydrogen citrate
E262	Sodium hydrogen diacetate
E450(a)	*di*Sodium dihydrogen diphosphate
621	Sodium hydrogen L-glutamate
350	Sodium hydrogen malate
E339(a)	Sodium dihydrogen orthophosphate
E339(a)	*di*Sodium hydrogen orthophosphate
E450(a)	*di*Sodium dihydrogen diphosphate
E222	Sodium hydrogen sulphite
524	Sodium hydroxide
631	Sodium inosinate
E325	Sodium lactate
350	Sodium malate
E223	Sodium metabisulphite
E219	Sodium methyl *para*-hydroxybenzoate
E251	Sodium nitrate
E250	Sodium nitrite
E232	Sodium orthophenylphenate
E339(c)	*tri*Sodium orthophosphate
E440(a)	Sodium pectate
E450(a)	*tri*Sodium diphosphate
E450(a)	*tetra*Sodium diphosphate
E450(b)	*penta*Sodium triphosphate
E450(c)	Sodium polyphosphates
E281	Sodium propionate
E217	Sodium propyl *para*-hydroxybenzoate

635	Sodium 5'-ribonucleotide
E470	Sodium salts of fatty acids
500	Sodium sesquicarbonate
E201	Sodium sorbate
E481	Sodium stearoyl-2-lactylate
514	Sodium sulphate
E221	Sodium sulphite
E335	*mono*Sodium L-(+)-tartrate
E335	*di*Sodium L-(+)-tartrate
E450(b)	*penta* Sodium triphosphate
E200	Sorbic acid
493	Sorbitan monolaurate
494	Sorbitan mono-oleate
495	Sorbitan monopalmitate
491	Sorbitan monostearate
492	Sorbitan tristearate
E420(i)	Sorbitol
E420(ii)	Sorbitol syrup
570	Stearic acid
E483	Stearyl tartrate
363	Succinic acid
E474	Sucroglycerides
E473	Sucrose esters of fatty acids
E220	Sulphur dioxide
513	Sulphuric acid
E110	Sunset yellow FCF
E307	Synthetic *alpha*-tocopherol
E308	Synthetic *beta*-tocopherol
E309	Synthetic *delta*-tocopherol
553(b)	Talc
E334	L-(+)-Tartaric acid
E472(d)	Tartaric acid esters of mono- and di-glycerides of food fatty acids
E102	Tartrazine
E450(a)	*tetra*Potassium diphosphate
E450(a)	*tetra*Sodium diphosphate
E233	Thiabendazole
E171	Titanium dioxide

E307	*alpha*-Tocopherol, synthetic
E309	*delta*-Tocopherol, synthetic
E308	*beta*-Tocopherol, synthetic
E306	Tocopherols, extracts of natural origin
E413	Tragacanth
380	*tri*Ammonium citrate
E333	*tri*Calcium citrate
E341(c)	*tri*Calcium diorthophosphate
E332	*tri*Potassium citrate
E340(c)	*tri*Potassium orthophosphate
E331(a)	*tri*Sodium citrate
E450(a)	*tri*Sodium diphosphate
E339(c)	*tri*Sodium orthophosphate
E153	Vegetable carbon
E161(e)	Violoxanthin
375	Vitamin B
E101	Vitamin B_2
E300	Vitamin C
E306	Vitamin E (natural)
E307	Vitamin E (synthetic)
E308	Vitamin E (synthetic)
E309	Vitamin E (synthetic)
907	Wax, refined, microcrystalline
E415	Xanthan gum
107	Yellow 2G

Glossary of Additive Terms

Acid

Acids are added to foods either to impart a sour or sharp flavour or for technological reasons, to control the level at which other substances in the food can function. The degree of acidity can thus be controlled in jams and preserves to regulate the optimum level of setting that can be achieved by the pectin in the fruit. Substances which neutralize acids are called alkalis or bases such as sodium, calcium and ammonium hydroxide, and substances which can hold the acid-alkali balance at a constant level are known as buffers.

Anti-caking Agents

These are substances which are added to foods such as icing sugar or salt or powdered milk to help them to flow freely and prevent the particles sticking together.

Antifoaming Agents

These are substances added to a food either to prevent excessive frothing on boiling or to reduce the formation of scum or to prevent boiling over. E900 Dimethylpolysiloxane, an inert silicone substance, is an example of an anti-foaming agent.

Antioxidants

Under normal circumstances fats and oils slowly become oxidized when they are exposed to the oxygen in the atmosphere. The process is accompanied by the development of a rancid 'off' flavour which if eaten can cause sickness. The addition of antioxidants to the fats prevents the process of oxidation. Antioxidants are also added to other non-fat foods such as cut fruits to prevent discolouration brought about by oxidation.

Artificial Sweeteners

These are substances, other than sugar, capable of producing a sweet taste.

Azo dyes

An azo dye has a particular chemical structure of the atoms in its molecule. It could be this 'azo' construction within the molecule to which a proportion of the population is sensitive, or it might be because of impurities. About a fifth of people who are sensitive to aspirin (usually middle-aged adults and more commonly women than men) are also sensitive to azo dyes. Other groups which may be affected are asthmatics and people who suffer from eczema.

The kinds of reactions that occur in sensitive people are contractions of the bronchi — the tubes allowing air into the lungs — (and asthmatic attacks), nettle rash, watering eyes and nose, blurred vision, swelling of the skin with fluid and in extreme cases shock and reduction in blood platelets with the production in the blood of anti-platelet antibodies. (The blood platelets are involved in blood clotting to seal wounds.)

It has been suggested by the late Dr Ben Feingold that azo dyes are among those substances which could trigger off the hyperactivity syndrome in children (see pages 12-14). The following are azo dyes:

E102   Tartrazine
E107   Yellow 2G
E110   Sunset yellow FCF
E122   Carmoisine
E123   Amaranth
E124   Ponceau 4R
E128   Red 2G
 154   Brown FK
 155   Chocolate brown HT

E151   Black PN
E180   Pigment rubine
'Coal tar dye' is an old name — it means nowadays that the dye is synthetically made and doesn't occur in nature. It would include all the above plus:
E104   Quinoline yellow
E127   Erythrosine
E131   Patent blue V
E132   Indigo carmine
E133   Brilliant blue FCF

Bases
Bases are added to foods to increase their alkalinity or reduce their acidity. Sometimes they are added to react with acids to give off carbon dioxide gas for aerating purposes.

Bleaching Agents
Substances employed to artificially bleach and whiten flour.

Buffers
Buffers are chemical substances which can resist considerable changes in the acid/alkali balance of solutions. The scale along which acid or alkali levels are measured is called the pH. Buffers (usually salts of weak acids) can maintain the pH at a predetermined level despite the addition of further acid or alkali.

Bulking Aids
Food additives which add to the bulk of the food but do not add to the calorific or energy value. The bulking aids are of value in slimming foods but they also help to 'pad out' or simulate more expensive ingredients.

Chelating Substances
When the acid/alkali ratio exceeds a particular limit or the ratio of traces of metal to one another exceeds a particular level, the trace metals may be precipitated out. The addition of a chelating substance such as EDTA (E385) retains the trace elements in the food solution, by bonding them on to an amino acid.

Coal Tar Dye
See azo dye.

Colour Index (C.I.) Numbers
These colour reference numbers are allocated in the Colour-Index of the Society of Dyers and Colourists (3rd Edition with 1975 revisions).

Colouring Matters
Water- or oil-soluble substances (or insoluble substances) which are produced artificially or are naturally occurring. Sometimes the colouring is permitted only on the outside of foods especially confectionery but usually it is permitted throughout the food.

Diluents
Substances which are used to dilute other additives or to dissolve them.

Emulsifiers
These are substances which can bring together oil, which is water-hating (hydrophobic) and water which is lypophobic (fat-hating) and mix them so that they do not separate out into layers. Some emulsifiers are plant gums, some are chemicals and others are synthetically produced derivatives of natural products.

Emulsifying Salts
A mixture of salts such as citrates, phosphates and tartrates which is added to cheese when it is melted as part of its processing to prevent the 'stringiness' which normally happens when cheese is cooked.

Excipients
This is normally a pharmacological term for 'inactive' powdered substances which are used to bind an 'active' drug into a tablet. The term is also applied in the baking industry to denote a carrier substance for additives used in bread.

Firming Agents

Calcium and magnesium salts are employed to retain the natural firmness or crispness of fruits and vegetables and to prevent their softening during the processing period.

Flavour Modifiers or Enhancers

These are substances used to enhance or reduce the taste or smell of a food without imparting any flavour of their own, so they are not 'flavours' as such, neither are they enzymes.

Gelling Agents

Substances which are capable of forming a jelly. Many of the gelling agents may be used in a stabilizing capacity too but not all stabilizers are capable of setting into a jelly.

Glazing Agents

Substances which either provide a shiny appearance or polish to the food or provide a protective coat, or both.

Humectants

These are substances which absorb water vapour from the atmosphere and prevent the food from drying out and becoming hard and unpalatable. Glycerine is added to royal icing in the home as a humectant, to prevent the icing drying out and hardening.

Liquid Freezants

Liquids or liquefiable gases which can freeze food by coming into contact with it directly and extracting heat from it.

Mineral Hydrocarbons

A wide variety of substances derived from bitumen (paraffin hydrocarbons) whether liquid, semi-liquid or solid. The group includes white oil, liquid paraffin, petroleum jelly, microcrystalline wax and hard paraffin.

Packaging Gases

These are inert gases which are employed to occupy space in packaging which if occupied by atmospheric air would cause oxidation of the contents or encourage the growth of micro-organisms.

Preservatives
Preservatives are substances which inhibit the growth of bacteria, fungi and viruses within foods and thus prevent the spoilage of these foods. Gases such as sulphur dioxide, organic and inorganic acids, phosphates and nitrates are all preservatives.

Propellants
Gases or volatile liquids employed in aerosol containers to expel the contents when the button is depressed.

Release Agents
Substances added to the machinery or coated onto food to prevent foods from sticking to the surfaces of food-processing equipment such as moulds, conveyors, cooking pans or trays. Release agents such as magnesium stearate are also added to tins in which foods are packaged to allow the contents to slip out easily.

Sequestrants
Traces of metals, always present in the environment, can cause deterioration in food by advancing the oxidation process, or cause premature setting in dessert mixes. Sequestrants are substances capable of attaching themselves to the trace metals such as calcium, iron or copper.

Solvents
These are liquids which are used to disperse substances either in solution or in suspension. They may also be used for extraction, then can either remain, as can be the case with the alcoholic extraction of a flavour, or be removed, as when oil is dissolved from seed.

Stabilizers
Similar in function to emulsifiers and thickeners, stabilizers serve to protect the droplets in an emulsion from collision with one another and consequently their tendency to separate out.

Stabilizers reduce coalescence either by adding to the viscosity or thickness of the medium forming a protection to the droplets or by forming protective colloids so the frequency and energy of collisions are minimized. The term 'stabilizer' may embrace thickening and gelling agents.

Synergists

A synergist is a substance which is capable of increasing or enhancing the effect of another substance. In the context of food additives, synergists are usually used to enhance the effects of antioxidants. These synergists include tartaric and citric acid and their calcium potassium and sodium salts.

Thickeners

Food additives which add to the viscosity of a food. Most of the thickeners employed are of plant origin, for example, seaweed or algae derivatives or substances produced from cellulose capable of forming a gel or colloid. Silicon dioxide (551) is the only inorganic additive employed as a thickener.

Vitamins

Most vitamins are not regarded as food additives unless they are fulfilling the function of a food additive such as that of antioxidant in the case of E300 L-Ascorbic acid (vitamin C) or E306-9 Tocopherols (vitamin E) or a colour in the case of 101a Riboflavin-5' phosphate (vitamin B_2).